Sauro Succi

An Introduction to Computational Physics
Part I: Grid Methods

APPUNTI

SCUOLA NORMALE SUPERIORE
2002

ISBN: 978-88-7642-263-8

Sauro Succi
Consiglio Nazionale delle Ricerche
Istituto per le Applicazioni del Calcolo
Viale del Policlinico, 137
I-00161 ROMA, Italy
succi@iac.rm.cnr.it

An Introduction to Computational Physics. Part I: Grid Methods

PREFACE

This volume collects the first part of the lecture notes of the course "An Introduction to Computational Physics" held in the academic year 2000-2001 for the students of the University of Pisa and Scuola Normale Superiore at the level of third and fourth year undergraduates in physics and chemistry. The lectures plan splits into two major themes:

1. *Grid methods*:
2. *Particle methods*:

Grid methods are the tool of the trade for the solution of ordinary and partial differential equations and consequently they represent a "must" for anyone dealing with computational science. Within grid methods, a major distinction is made between methods which do not require matrix algebra and those which do.

1. Matrix-Free methods

Matrix-free methods are usually associated with *explicit* schemes. Explicit schemes are numerical methods which comply with the basic principles of *locality* and *causality*. In these methods the actual state of the system at a given space location depends on the immediate past of a local spatial neighborhood. Preserving causality and locality rules out action at distance, hence it does not introduce any simultaneous relation between the various unknowns. This is why no matrix algebra is (generally) required. Three major real-space techniques: Finite differences, Finite Volumes, Finite Elements are presented, which correpond to an increasing degree of geometrical complexity. In addition a prominent reciprocal-space technique, the Spectral method, is also discussed.

2. Matrix methods

Once causality is abandoned (to march in large steps in time) systems of simultaneous equations result, which require the use of matrix algebra regardless of the type of space-differencing method adopted. The solution of such systems of equations is a computationally demanding task for which a vaste repertoire of techniques is available from modern Numerical Analysis. We shall provide a cursory view of the basic methods to solve large systems of linear (and non-linear equations) both in direct and iterative form. In addition, a cursory view of the main techniques to solve eigenvalue problems is provided.

Acknowledgments

I wish to express my deep gratitude to Prof. Mario Tosi, whose kind support, advice and encouragement all along, made this project possible altogether. I would also like to thank Dr Marilu' Chiofalo for several useful discussions on the subjects of this book, and Dr A. Sportiello for his critical help with the preparation of this manuscript.

CONTENTS

1 General properties of numerical schemes   5
  1.1   The continuum problem   5
  1.2   Living in a discrete world   6
     1.2.1   Consistency   6
     1.2.2   Accuracy   8
     1.2.3   Stability   8
     1.2.4   Efficiency   9
     1.2.5   Optimization   10
  1.3   References   11

2 Time marching   13
  2.1   Explicit versus implicit time-marching   13
     2.1.1   Dispersion relations   14
     2.1.2   Stability of implicit schemes   16
     2.1.3   Non-linear problems   16
  2.2   Higher order schemes: Runge-Kutta   17
     2.2.1   ERK's: Explicit Runge-Kutta   18
     2.2.2   IRK's: Implicit Runge-Kutta   19
  2.3   Second order derivatives   20
     2.3.1   Leapfrog marching   20
  2.4   References   21

3 Finite Difference Method   23
  3.1   The advection equation   23
  3.2   Centered spatial differences   24
     3.2.1   Upwind method   25
     3.2.2   Lax scheme   26
     3.2.3   Lax-Wendroff method   27
  3.3   Numerical renormalization   28
     3.3.1   Operator identities and "numerical renormalization"   28
  3.4   Other equations   29
     3.4.1   Advection-Diffusion   29
     3.4.2   The Klein-Gordon equation   30
     3.4.3   The Dirac equation   30
  3.5   Boundary Conditions   31
  3.6   References   33
  3.7   Projects   33
  3.8   Warm-up programs   33

2

4   Finite Volume Method   35
   4.1   Finite Volume formulation   35
   4.2   Discrete Equations of motion   37
   4.3   Time discretization   38
   4.4   Topological constraints: how far can we go?   38
   4.5   Computational complexity   39
   4.6   Projects   39
   4.7   References   40

5   Finite Element Method   41
   5.1   Finite-Element formulation   41
   5.2   The equations of motion: projection and weak solu-
          tions   43
   5.3   1D FEM: Hat functions   44
          5.3.1   FEM in 2D: The triangle   45
   5.4   Matrix Assembly   45
   5.5   Boundary conditions   47
   5.6   Computational complexity   47
   5.7   FEM for Quantum Field Theory   47
   5.8   FEM for quantum gravity: Regge calculus   48
   5.9   References   49
   5.10 Projects   49

6   Spectral Method   51
   6.1   Fourier and Discrete Fourier transforms   51
   6.2   The magic FFT   53
   6.3   Pseudo-spectral method   56
          6.3.1   Operator-splitting methods   57
   6.4   Computational efficiency   57
   6.5   References   57
   6.6   Projects   57
   6.7   Warm-up program   57

7   Direct Methods   65
   7.1   Basic matrix algebra   65
   7.2   Gaussian elimination   66
          7.2.1   The LU factorization   69
   7.3   Gaussian elimination in 1D: the Thomas method   69
   7.4   Software considerations   70
   7.5   References   70
   7.6   Projects   70

8   Iterative methods   71
   8.1   Generalities   71
   8.2   Theoretical Conditions for convergence   72
   8.3   Convergence criteria   73

8.4	Operator Splitting methods	74
	8.4.1 Jacobi splitting	74
	8.4.2 Gauss-Seidel splitting	75
	8.4.3 Successive Under and Over-Relaxation	75
	8.4.4 Tsebitchev acceleration	76
	8.4.5 Block-SOR/SUR iterations	76
8.5	Dynamic relaxation methods	77
8.6	Gradient methods	78
8.7	Conjugate-Gradient methods	79
8.8	Preconditioning	81
8.9	Multigrid methods	82
	8.9.1 Multigrid formulation	82
	8.9.2 Fine-to-Coarse: Restriction	83
	8.9.3 Coarse-to-Fine: Extension	83
	8.9.4 Multigrid implementation	84
	8.9.5 Multigrid schedule	85
8.10	Iterative or Direct?	86
8.11	References	86
8.12	Projects	86

9 Systems of non linear equations — 87
9.1	General problem	87
9.2	Fixed-point methods	87
9.3	Newton-Raphson method	88
9.4	Dynamic relaxation methods	90
	9.4.1 Deterministic dynamics	90
	9.4.2 Stochastic (Langevin) dynamics	91
9.5	Genetic methods	91
9.6	Projects	92
9.7	References	92

10 Eigenvalue Problems — 93
10.1	General problem	93
10.2	Mathematical problem	94
10.3	Direct eigenvalue solvers: Housolder rotations	95
	10.3.1 Eigenvalues of tridiagonal matrices	96
10.4	Inverse iteration	97
	10.4.1 Direct iteration	97
	10.4.2 Inverse iteration	98
10.5	Monster matrices: the Lanczos method	99
	10.5.1 Krylov subspaces and Lanczos method	99
	10.5.2 The Lanczos algorithm	100
10.6	Dynamic minimization methods	101
	10.6.1 Dissipative dynamics	101
	10.6.2 Conservative dynamics	102

4

10.7 Projects   102
10.8 References   102

1

GENERAL PROPERTIES OF NUMERICAL SCHEMES

In this chapter we shall present a cursory view of the general properties which govern the correct discretization of a given continuum problem expressed in the form of ordinary/partial differential equations. These properties are strongly tied to the basic principles (symmetries, conservation laws, evolutionary constraints) which lie at the heart of the physical model. The golden rule to a healthy transcription of the continuum problem to the discrete world is to preserve the integrity of these basic principles.

1.1 The continuum problem

For the sake of definiteness, we shall make reference to the following evolutionary problem:

$$\partial_t \Psi = L\Psi,$$
$$\Psi(x, t = 0) = \Psi_0(x) \tag{1.1}$$

supplemented with boundary conditions on the boundary $\partial\Omega$ of the spatial domain Ω where the space variable x is defined. Here L is the Liouville operator (not necessarily linear) generating the time evolution of the system described by the state function $\Psi(x, t)$, eventually a vector function. This abstract notation encompasses a number of important linear and non-linear equations:

$$
\begin{aligned}
&L = U\partial_x, &&\text{Advection} \\
&L = D\partial_{xx}, &&\text{Diffusion} \\
&L = -iD\partial_{xx} + iV(x), &&\text{Schroedinger} \\
&L = -\Psi\partial_x, &&\text{Self-convection (Burgers)} \\
&L = -\Psi\partial_x + K\partial_{xxx}, &&\text{(Korteweg-de-Vries)}
\end{aligned}
\tag{1.2}
$$

to cite just a few. The formal solution of eq. (1.1) reads as follows:

$$\Psi(t) = e^{Lt}\,\Psi(0) \equiv T_t\,\Psi(0) \tag{1.3}$$

and defines the *resolvent* operator, or *time propagator*, as:

$$T_t \equiv e^{Lt} \tag{1.4}$$

This operator fulfills the (semi)-group properties:

$$T_{t+u} = T_t \cdot T_u, \qquad\qquad t, u > 0 \qquad\qquad (1.5)$$

$$T_0 = I \qquad\qquad\qquad (1.6)$$

$$T_{-t} \cdot T_t = I, \qquad\qquad \text{(reversible systems only)} \qquad (1.7)$$

which must hold independently of its specific form. Computational physics is the 'art' of controlling the 'distorsions' suffered by this operator (and the corresponding solution) as space-time is made discrete. Let us therefore move on to the basics of discretization.

1.2   Living in a discrete world

The starting point of any discretization technique is to replace continuum space-time C with a discrete spacetime C_h, where $h \equiv (dt, dx)$ is the 'quantum' of granularity of discrete space-time. For simplicity we shall often refer to one-dimensional systems (1+1 dimensions in space time) living on a segment $[0, L]$ over a time span $[0, T]$. The corresponding discrete space-time is a uniform discrete lattice:

$$x_l = l \cdot dx, \qquad l = 0, \dots, N_x, \qquad L = (N_x - 1) \cdot dx \qquad (1.8)$$

$$t_n = n \cdot dt, \qquad n = 0, \dots, N_t, \qquad T = (N_t - 1) \cdot dt \qquad (1.9)$$

The time propagator can be expressed as the sum of a discrete propagator T_h plus and an *error operator* E_h:

$$T = T_h + E_h \qquad\qquad (1.10)$$

and similarly for the solution:

$$\Psi = \Psi_h + e_h \qquad\qquad (1.11)$$

Our task is to analyze the "physical distorsions" associated with the error operator E_h, depending of the type of numerical discretization adopted. We should make sure to avoid "computer hallucinations", spurious physics generated by E_h!!! In particular, we shall examine the following CASE properties:

- *Consistency*
- *Accuracy*
- *Stability*
- *Efficiency*

1.2.1   *Consistency*

Consistency requires the discrete operators to recover the continuums one when the lattice quantum h is sent to zero:

$$T_h \to T \quad \text{as} \quad h \to 0, \qquad\qquad (1.12)$$

$$E_h \to E_0 = 0 \quad \text{as} \quad h \to 0, \qquad\qquad (1.13)$$

Whenever $E_0 \neq 0$ we shall say that the numerical scheme is affected by an *anomaly*, typically a badly broken symmetry of the Liouville equation (Lorenz invariance, time inversion, parity and so on). Of course, the limit $h \to 0$ must be intended in the sense of 'phase-transitions', namely 'vanishingly small' as compared to any lengthscale of physical interest, not strictly $h = 0$. Typical cases of inconsistency arise when the discrete operator does not preserve the basic symmetries of the continuum. For example, if the Liouville operator entails, say, first order space derivatives only, a second order approximation of the space derivative and a first order approaximation of the time derivative are often at risk of generating inconsistencies. A useful way of helping consistency is to guarantee that prime integrals of the Liouville equation, i.e. invariants of motion, remain such after discretization. In other words, let $\phi(x)$ a given function such that

$$\int \phi(x) L\Psi(x,t) dx = 0$$

then, by (1.1), the integral quantity:

$$I(t) = \int \phi(x) \Psi(x,t) dx$$

is an invariant of motion, namely:

$$dI(t)/dt = \int \phi L\Psi dx = 0$$

A typical example is the total probability $P(t) = \int \psi^*(x,t)\psi(x,t)dx$ in the Schroedinger equation. Invariants of motion play a paramount role in physics and therefore any good numerical scheme must handle them with due respect:

Physical invariants must turn into numerical invariants! This does not mean that the actual value of numerical prime integral, say I_h, must be the same as its continuum continuum counterpart, I. In general it is not. The key requirement, however, is that such quantity, whatever its value, remains *exactly* the same (up to machine roundoff) in the course of the numerical time evolution:

$$\frac{|I_h(t) - I_h(0)|}{|I_h(0)|} < \epsilon$$

This property, sometimes also called *Conservativeness* is a a powerful medicine against numerical instability.

In a broad sense, Conservativeness is in charge of securing compliance with the first principle of thermodynamics. For *dissipative* systems an equally important property is *Dissipativeness*, namely compliance with the second principle of thermodynamics. In essence, conservativeness means that quantities which evolve monotonically in time (Lyapunov functions)

$$d\Lambda/dt \geq 0$$

should do the same after discretization.

A prototypical Lyapunov functional is the famous Boltzmann H-function $H(t) = \int f \ln f \, dv \, dx$ where $f \equiv f(x,v,t)$ is the probability of finding a molecule in x at time t with speed v. After a sign change, the H-function delivers the entropy of the gas of molecules. It is quite natural that a good numerical scheme for a dissipative system should guarantee that the numerical entropy is a monotonically non-decreasing function of time.

1.2.2 Accuracy

Accuracy refers to the extent to which T_h is a 'good' approximation of T. In general we say that a given discretized operator O_h is p-th order accurate if it does not produce any error as applied to a polynomial of degree p. Symbolically:

$$E_h = O - O_h \sim h^p \tag{1.14}$$

We shall see several practical examples in the sequel, mostly in connection with differential operators such as d/dx and similar. For instance, it is readily checked that the simple first order forward difference operator $[f(x+h)-f(x)]/h$ is only first order accurate, while the centered operator $[f(x+h)-f(x-h)]/2h$ is second order.

Accuracy is obviously very important since it permits to save on the number of grid points needed to achieve a given level of accuracy. Let $e_k = C_k h^k$, $k = 1, 2$ the errors generated by a first and second order accurate operators respectively. At a given, prescribed, level of accuracy, the grid sizes h_k are related as follows:

$$h_2 = \sqrt{\frac{C_1}{C_2}} h_1$$

Iff the prefactors C_1, C_2 are of the same order of magnitude the above relation shows that $h_2 \sim \sqrt{h_1}$ and since they are both small quantities, much smaller than one (in lattice units), the result is $h_2 \gg h_1$, which means many less grid points $N_2 = L/h_2 \ll N_1 = L/h_1$. In practice the condition $C_2 \sim C_1$ must be watched carefully: a big prefactor can spoil the advantage of higher accuracy!

1.2.3 Stability

A numerical scheme is said to be stable if the error in the initial conditions does not grow indefinitely in time. Actually, if it stays below a prescribed tolerance ϵ.

$$\lim_{t \to \infty} \frac{e(t)}{e(0)} < \epsilon \tag{1.15}$$

The prototypical case is the simple decay equation:

$$\frac{d\psi}{dt} = \omega\psi$$

with $\omega < 0$. A simple first order time marching scheme yields:

$$\psi(t + dt) = (1 + \omega dt)\psi$$

from which it is easily seen that numerical stability implies the following restriction:

$$-2 < \omega dt < 0$$

namely,

$$dt < 2/|\omega|$$

A similar statement applies to the general framework of the Liouville equation, where the stability condition given above applies to the entire spectrum of the (discrete) Liouville operator, and notably, to the largest non-zero eigenvalue, which is in direct control of the numerical time-step.

1.2.4 *Efficiency*

Every numerical scheme is destined to computer implementation. The corresponding computer *algorithm* can be characterized by its *computational complexity*, roughly speaking the number of operations required to complete the task of evolving the system in time up to a given time T. Computational complexity is affected by several factors, primarily the number of degrees of freedom DOF, which in turn scale with the number of spacetime points $N = N_x \cdot N_t$:

$$DOF \sim C(N)N \qquad (1.16)$$

where C is the number of operations per grid point (computational density). In general, the computational density scales with a power of the total number of grid points:

$$C \sim N^\xi \qquad (1.17)$$

where ξ is a complexity exponent which depends on the numerical scheme adopted. When the numerical scheme is *local*, i.e. the state at a given site depends

FIG. 1.1. *Geometrical picture of stability.*

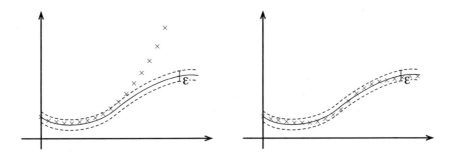

only on a limited number of neighbors, then $C = const.$. Global schemes are instead characterized by $\xi > 0$.

From the practical point if view efficiency relates not only to attaining the least number of operations (Computational density). but also to the quality of operations $(\pm, *, /)$, divisions being generally 3-4 times more expensive.

The processor speed is measured in Mflop/s (Millions floating point/second) For instance, a computer with a $100MHz$ clock-rate provides 1 flop/cycle, hence it yields 100 Mflops/s.

It is very important to keep in mind that *data access* also takes time. To appreciate this point, we recall that computer memory is generally organized in hierarchical way:

It is therefore important to work with local data: whenever data has been accessed from central memory and loaded into local cache memory, it must be used as much as possible before being released back to central memory (cache reusability principle).

1.2.5 *Optimization*

The four CASE requirements are hardly disentangled, since each of them affects each other, often in a conflicting scenario. For instance, accurate high-order schemes require less grid-points than low-order ones. However, they also entail more computational work per grid point, so that it is not obvious a priori which one is the best. In addition, sophisticated discretizations are often more fragile in the sense of stability. In fact, the choice of the optimal scheme is often an art which requires not only knowledge of numerical analysis but also a good deal of experience and physical intuition.

This is why

Computational physics is not a subfield of numerical analysis!

In fact we shall invoke the following conjecture:

FIG. 1.2. *CPU registers.*

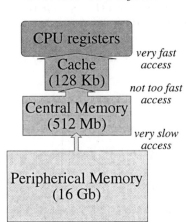

Numerical Correspondence Principle:

"Every numerical error can be interpreted as a fictitious physical process".

This is of course a mere conjecture, not a principle, but we shall learn that whenever numerical errors can be given a physical interpretation, we are already half-way towards the cure.

After these rather abstract notions it is time to take a closer look into the discretized world of computational physics. We shall begin by considering the time variable, the subject of the next chapter.

1.3   References

1. D. Potter, Computational Physics, Wiley, 1976

2

TIME MARCHING

In this lecture we shall provide the main guidelines for the discretization of the time variable. This is especially important since it bears upon the principle of causality and locality. As we shall see, completely distinct classes of methods arise depending on whether or not such principle is reflected in the discrete-time formulation.

2.1 Explicit versus implicit time-marching

Owing to the one-sided, causal, nature of the time variable, a friutful starting point for constructing discrete time marching scheme is to integrate forward in time over a 'small' interval t to $t + dt$. A simple quadrature delivers:

$$\Psi(t + dt) = \Psi(t) + \int_t^{t+dt} L\Psi(t')\, dt' \qquad (2.1)$$

The left-hand side is exact and therefore the numerical approximation enters only in the way the integral at the right hand side is handled. Let us for now assume for simplicity that the operator L is linear. Using a two-point trapezoidal rule, we obtain

$$\Psi(t + dt) = \Psi(t) + dt[aL\Psi(t) + bL\Psi(t + dt)] \qquad (2.2)$$

where $0 < a < 1$ is the weight of 'present' state at time t and $b = 1 - a$ the weight of the future state $t + dt$ (anticipation parameter).

Three important choices stand out:

- Fully Explicit: $a = 1, b = 0$, No anticipation
- Fully Implicit: $a = 0, b = 1$, Full anticipation
- Cranck-Nicolson: $a = b = 1/2$, Half anticipation

The above expression shows that any *any degree of anticipation is bound to break causality*. Breaking causality means loosing locality, that is, in computational terms, solving matrix problems. This brings in a crucial distinction between *Explicit* and *Implicit* methods. It is useful to recast eq. (2.2) in matrix form:

$$A\Psi(t + dt) = B\Psi(t) \qquad (2.3)$$

where:

$$A = I - bLdt$$
$$B = I + aLdt \qquad (2.4)$$

where I is the identity matrix. This identifies the time propagator: $(h \equiv dt)$

$$T_h = A^{-1}B \qquad (2.5)$$

From this expression we draw a basic conclusion:

Explicit methods do not require matrix algebra.

In fact, if $a = 0$ the matrix A is merely the identity. On the other hand, *any* degree of anticipation involves the solution of a matrix problem. This is why the Explicit/Implicit distinction gives rise to entirely different classes of numerical methods.

In this lecture we shall be concerned with the former.

2.1.1 *Dispersion relations*

The main properties of the discrete propagator T_h are most conveniently analysed in terms of *dispersion relations*. According to the Laplace representation of the resolvent operator:

$$e^{Lt} = \frac{1}{2\pi i} \int_{C-i\infty}^{C+\infty} \frac{e^{tz}}{L-z} dz \qquad (2.6)$$

we introduce a complex variable z whih takes on values in operator spaces. The real part of z is associated with monotonic growth/decay of the solution, whereas the imaginary one describes periodic oscillations.

The discrete propagator associated to (2.4) reads:

$$T_h(z) = \frac{1 + az}{1 - bz} \qquad (2.7)$$

It is immediately seen that (2.7) is a *rational approximant* (Pade' approximant) of the exact propagator $T(z) = e^z$. The specific Pade's approximants corresponding to Fully Explicit, Cranck-Nicolson and Fully Implicit are given below:

$$
\begin{aligned}
P_{1,0} &= (1 + z) \\
P_{1,1} &= (1 + z/2)/(1 - z/2) \\
P_{0,1} &= 1/(1 - z)
\end{aligned}
\qquad (2.8)
$$

It is interesting to observe that all of these approximants are *conformal transformations* of the z plane. In fact, the representation 2.7 is a Mobius transformation $z' = \frac{az+b}{cz+d}$ with the conformally invariant property $ad - bc = 1$.

To analyse the dispersion relation associated with time marching relation 2.3, we Fourier-transform $\Psi \sim e^{-i\omega t}$. This delivers:

$$e^{-i\omega h} = T_h(z) \qquad (2.9)$$

Since T_h contains space derivatives, Fourier transformation in space turns the Liouville operator into a function of the wavenumber $k = -id/dx$, $z = z(k)$, so that the above relation is converted into a *numerical dispersion relation*:

$$e^{-i\omega h} = T_h[z(kd)] \tag{2.10}$$

where d is the lattice spacing in configuration space. This equation contains all the details of *interaction with the lattice*. In fact we can regard d, h as *coupling strengths* to space-time discreteness. We then expect numerics to produce typical "solid-state" physical phenomena, such as instabilities, diffusion, dispersion and others.

By letting $\omega = \Omega + i\gamma$, the dispersion relation associated with (2.3) is:

$$\Omega_h = \ln \frac{|T_h|}{h} \tag{2.11a}$$

$$\gamma_h = \frac{\arctan \frac{\Im T_h}{\Re T_h}}{h} \tag{2.11b}$$

where $\Re T$, $\Im T$ represent the real and imaginary parts of T respectively.

In the limit $h \to 0$ the above relations are supposed to recover the exact disperison relation in the continuum. The mathematical statement of the Numerical Correspondence Principle (see previous lecture) is that these errors always come in the form of sensible combinations of the discrete time step h and mesh size d into physically meaningful dimensionless numbers.

For instance, for pure advection at speed U, we shall meet the famous Courant number:

$$C = \frac{Uh}{d} \tag{2.12}$$

representing the ratio of physical speed, U, to the lattice speed $U_g = d/h$. Each physical phenomenon gives rise to its own Courant number and we shall learn that $C < C_{\max} < 1$ is a general condition for the stability of explicit schemes. We shall see specific instances in next chapter, depending of the type of Liouville operator and discretization scheme. However, a number of important conclusions can be drawn without entering the details of the Liouville operator. For instance, from the numerical dispersion relation we immediately conclude that Stability imposes the following constraint:

$$|T_h| < 1 \tag{2.13}$$

In other words, the propagator must be a *contractive map*. Based on the expression 2.7 for the explicit scheme, this means

$$|z| < 1 \tag{2.14}$$

This is fulfilled in the strip

$$-2 < x < 0 \tag{2.15}$$

where $x = Real(z)$, and translates into a stringent limitation the time-step, generally known as *Courant-Friedrichs-Lewy* (CFL) conditions.

These conditions respond to the idea that *no physical process can proceed faster than allowed by the discrete size of the lattice.*

2.1.2 *Stability of implicit schemes*

Inspection of implicit propagators, say Cranck-Nicolson, shows that the stability condition is fulfilled for *any* value of z since $|1 - z/2/(1+z/2| < 1$ for any $x < 0$.

In other words, breaking causality by anticipation of the future buys us stability, regardless of the time-step size: no more CFL restrictions!

The price of stability is non-locality: even though the Liouville operator is local, the inverse of $(1 + Ldt/2)$ is not. In fact, by expanding the denominator we obtain an infinite series of powers $1 + z + z^2/2 + ...z^p/p!$, so that each, say, first order spatial derivative in L generates higher derivatives of *all* orders! In fact, since $z = z(k)$ and k is a local space-connector (first order derivative) the generic monomial z^p corresponds to a simultaneous connection between sites p sites apart in the spatial lattice.

Does this mean we should always go for implicit time marching? Of course not!

First of all, implicit time marching implies the solution of a linear algebraic problem, which is a costly task. Second, large time-steps might hit the numerical accuracy.

In general, implicit schemes are recommended for problems where the main focus is on steady-state solutions rather than dynamical evolutions.

2.1.3 *Non-linear problems*

The analysis so far was restricted to the case of linear Liouvillean operators L. If L is non-linear, the treatment becomes obviously more complicated.

$$\hat{\Psi} = \Psi + h \left[aL(\Psi) + bL(\hat{\Psi}) \right] \qquad (2.16)$$

where hat means at $t + dt$. For explicit schemes ($b = 0$), the above formulation still remains matrix-free. For implicit schemes however, we are confronted with

FIG. 2.1. *Stability diagrams in the complex plane: Fully explicit.*

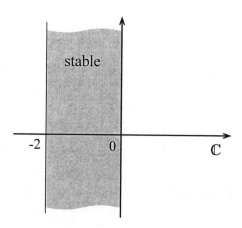

FIG. 2.2. *Fully implicit, and also Cranck-Nicholson.*

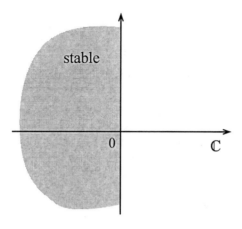

FIG. 2.3. *Spacetime diagram of implicit schemes: dt is much larger than dx/U.*

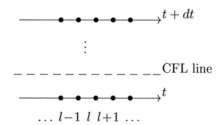

a *non-linear algebraic* problem since the right hand side depends (non-linearly) on the actual unknown. These problemis are typically handled by functional iterative techniques, to be shortly described in a future chapter.

2.2 Higher order schemes: Runge-Kutta

The accuracy of the previous methods can be enhanced by using higher order approximations of the time integral on the right hand side of (2.1). This generates so-called *multilevel* or *multistep* procedures, such as Runge-Kutta and others.

The idea is to perform the numerical integration of the right hand side of the eq. (2.1) using a multi-point quadrature. With, say, K time-levels, and letting $\Phi(\Psi) \equiv L\psi$, this yields:

$$\Psi(t + dt) - \Psi(t) = \sum_{k=0}^{K-1} w_k \Phi[\Psi(t + \tau_k), t + \tau_k] \qquad (2.17)$$

where

$$\tau_k = c_k dt$$

w_k and c_k being a set of weights and nodes specific to the quadrature rule adopted. An optimal choice of these quadrature parameters, such as Gaussian quadrature, is known to provide order h^{2K} accuracy, potentially much better than the first/second orders schemes discussed thus far. The problem is, however, that the intermediate values Ψ_k are not available, so that a procedure is needed to compute them.

2.2.1 ERK's: Explicit Runge-Kutta

The idea behind Runge-Kutta schemes is to proceed through a sequence of K intermediate stages and compute the of Ψ_k by using values at previous sub-steps. Denoting by:

$$y_k \equiv \Psi(t + \tau_k) - \Psi(t)$$

the increment at step k, a typical K-step *explicit* Runge-Kutta sequence looks like follows:

$$y_0 = 0 \tag{2.18a}$$
$$y_1 = dt R_{10} \Phi_0 \tag{2.18b}$$
$$y_2 = dt(R_{20}\Phi_0 + R_{21}\Phi_1) \tag{2.18c}$$

$$\cdots$$

$$y_{K-1} = dt \sum_{k=0}^{K-1} R_{Kk}\Phi_k \tag{2.18d}$$

where

$$\Phi_k = \Phi(t + \tau_k, \Psi(t) + y_k)$$

Once the transfer matrix R_{kj} is specified the above scheme permits to compute the sequence of values y_k, hence Ψ_k needed to proceed up to the final stage $t + dt$. The question is: how to compute the elements of the low-triangular transfer matrix R? Essentially by imposing that the RK scheme solves exactly the original equation up to polynomials of order K.

A practical example will best illustrate the point.

Consider the case $K = 2$:

$$y_0 = 0, \qquad\qquad \Phi_0 = \Phi(\Psi) \tag{2.19}$$
$$y_1 = dt R_{10}\Phi_0, \qquad\qquad \Psi_1 = \Psi(t) + y_1 \tag{2.20}$$

and finally:

$$\Psi(t + dt) = \Psi(t) + h[w_0\Psi_0 + w_1\Psi_1] \tag{2.21}$$

FIG. 2.4. *A 3-stage ERK: Explicit Runge Kutta scheme. All connectors R_{kj} point forward in time according to the causal, explicit, character of the scheme.*

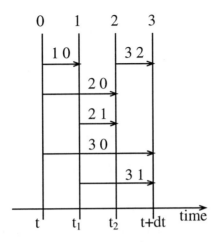

By Taylor expanding the above espression to first order in h we obtain

$$\Psi(t+dt) = \Psi(t) + h[w_0 + w_1]\Psi_0 + hw_1[\Phi_y(t)y_1 + \Phi_t(t)\tau_1] \qquad (2.22)$$

By replacing the expressions of y_1 and τ_1, this becomes:

$$\Psi(t+dt) = \Psi(t) + h[w_0 + w_1]\Psi_0 + h^2 w_1[R_{10}\Phi(t)\Phi_y(t)y_1 + c_1\Phi_t(t)] \qquad (2.23)$$

This is requested to match a second order Taylor expansion of the original solution:

$$\Psi(t+dt) = \Psi(t) + h\Psi_t(t) + \frac{h^2}{2}\Psi_{tt}(t) \qquad (2.24)$$

By using the identity $\Psi_{tt} = \Phi_t + \Phi\frac{\partial\Phi}{\partial\Psi}$ equating 2.23 and 2.24 delivers the following relations

$$\begin{aligned} w_0 + w_1 &= 1 \\ w_1 R_{10} &= 1/2 \\ w_1 c_1 &= 1/2 \end{aligned} \qquad (2.25)$$

which completely specify the explicit Runge-Kutta algorithm. Higher order schemes can be constructed along the same lines, although it should be clear that algebraic complexity grows exponentially with the order of the scheme. In practice, it is very rare to go beyond $K = 4$.

2.2.2 IRK's: Implicit Runge-Kutta

From the above discussion it is clear that the RK scheme can be made implicit by letting the transfer matrix R_{ik} carry backward (future-to-past) dependencies,

$R_{ik} \neq 0$ also for $k > i$. IRK's allow larger timesteps but they tend however to be fairly complicated since at each timestep (and lattice site!) we now have a K dimensional non-linear algebraic problem to solve. In general, care needs be exercised with such higher order schemes because: i) they are memory and time consuming, ii) they are more exposed to numerical instabilities. In depth treatment of this subject can be found in the huge literature of ordinary differential equations.

2.3   Second order derivatives

So far we have considered first order time evolutionary phenomena. A huge variety of natural systems, and notably wave-propagation phenomena, obey equations with second order time derivatives ($F = ma$ in a generalized sense): Neglecting space derivatives for simplicity, we have:

$$\partial_{tt}\Psi = f(\Psi) \tag{2.26}$$

We can stick to a first-order time formalism provided that the state vector is expressed as coordinate-momentum decomposition $\Psi(t) = [q(t), p(t)]$: The eq. (2.26) takes then the first-order form:

$$\begin{cases} \dfrac{dq}{dt} = p \\[2mm] \dfrac{dp}{dt} = f(q) \end{cases} \tag{2.27}$$

where spece-dependence has been omitted since we concentrate on time-stepping alone.

2.3.1 *Leapfrog marching*

For second order systems an imperative property is *reversibility*: we should be able to roll the scheme back in time $t \to -t$ without loosing information. A very useful time-stepper for 2nd order derivatives is the Leapfrog method:

$$q^{n+1} - q^n = p^{n+1/2} \cdot dt \tag{2.28}$$
$$p^{n+3/2} - p^{n+1/2} = f^{n+1} \cdot dt \tag{2.29}$$

This corresponds to a staggered time marching:

$$q_0 \qquad q_1 \qquad q_2 \qquad q_3 \qquad q_4 \quad \cdots$$

$$\text{time}$$

$$p_{\frac{1}{2}} \qquad p_{1+\frac{1}{2}} \quad p_{2+\frac{1}{2}} \quad p_{3+\frac{1}{2}} \quad p_{4+\frac{1}{2}} \quad \cdots$$

The Leapfrog is easily shown to be equivalent to a all-q time marching scheme:

$$q^{n+1} - 2q^n dt + q^{n-1} = f^n dt^2 \tag{2.30}$$

which is a simple-minded fourth-order accurate representation of the d^2/dt^2 operator. Leapfrog schemes are a popular choice for evolutionary hyperbolic PDE's, such as:

$$\phi_{tt} - c^2 \phi_{xx} = f[\phi] \tag{2.31}$$

and generally speaking for linear and non-linear wavelike problems.

We shall cover this subject in more depth when dealing with the solution of large systems of ordinary differential equations, in second part of this course.

2.4 References

- D. Potter, Computational Physics, Wiley and Sons, 1976
- A. Iserles, Numerical analysis of differential equations, Cambridge Univ. Press, 1996

3

FINITE DIFFERENCE METHOD

In this chapter we shall deal with discretization of the space variable and related differential operators.

A few major finite-difference schemes are presented, together with the corresponding numerical dispersion relations. The important concepts of numerical diffusion and dispersion are illustrated with specific examples arising from the aforementioned finite-difference schemes.

3.1 The advection equation

The simplest, and most intuitive discretization technique for spatial variables is the finite-difference (FD) method. For simplicity, we shall again refer to a uniform regular lattice ("perfect crystal") of uniform spacing d, and lattice coordinates $x_l = l \cdot d, l = 1, L$.

FIG. 3.1. *A uniform lattice with L points.*

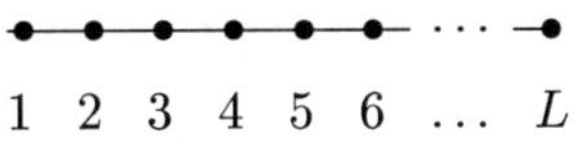

$$1 \quad 2 \quad 3 \quad 4 \quad 5 \quad 6 \quad \dots \quad L$$

We shall use the advection equation as a didactical vehicle to introduce a number of basic concepts. The advection equation is a paradigm of hyperbolic systems:

$$\partial_t f + U \partial_x f = 0 \tag{3.1}$$
$$f(x, 0) = f_0(x) \tag{3.2}$$

where $U = $ const. for simplicity.

The exact solution is the travelling wave:

$$f(x, t) = f_0(x - Ut) \tag{3.3}$$

corresponding to the dispersion relation:

$$\Omega = kU, \qquad\qquad \gamma = 0 \tag{3.4}$$

where $\omega = \Omega + i\gamma$:

The FD scheme should be designed to generate minimal deviations from this dispersion relation. Numerical errors on the real part of the dispersion relation are called "phase-errors" and generate *numerical dispersion* whereas errors on the imaginary part are referred to as 'amplitude errors' and generate *numerical diffusion*. Numerical diffusion with a negative diffusion coefficient is tantamount to instability.

3.2 Centered spatial differences

A simple and intuitive way of discretizing the advection equation is to use a centered difference for the spatial derivative

$$\frac{df}{dx} \sim \frac{f_{l+1} - f_{l-1}}{2d} \tag{3.5}$$

and a forward-difference for the time derivative. This yields:

$$f_l^{n+1} - f_l^n + \frac{c}{2}[f_{l+1}^n - f_{l-1}^n] = 0 \tag{3.6}$$

where n labels discrete time $t_n = n \cdot h$, and

$$c = Uh/d$$

is the Courant number.

In explicit form:

$$f_l^{n+1} = \frac{c}{2} f_l^{n-1} + f_l^n - \frac{c}{2} f_{l+1}^n \tag{3.7}$$

This scheme is 2nd order in space, first order in time, but virtually ... useless! Why? Because it is *unconditionally unstable*.

This is easily seen by inspecting its discrete dispersion relation:

$$\Omega = \arctan[c \sin kd]/h \tag{3.8}$$

$$\gamma = \frac{\ln[1 + c^2 \sin^2(kd)]}{2h} \tag{3.9}$$

The second of these equations immediately shows that $\gamma > 0$ for *any* finite value of kd and c, which means unconditional instability!

Where does such bad behaviour come from?

The reason is that the centered scheme breaks spacetime balance (covariance) of the original equation. The original equation has first order derivatives in both space and time, while the discretized one is 2nd accurate in space and only 1st order in time. Space is, so to say, overdiscretized. Consequently the discrete dispersion relation delivers a spurious mode, which turns out to be always unstable. Instability can also be given a very intuitive explanation in terms of particle motion. The scheme 3.7 can be interpreted as follows: the left neighbor contributes a fraction $c/2$ and the right neighbor a fraction $c/2$ to the central

FIG. 3.2. *Centered spatial differences.*

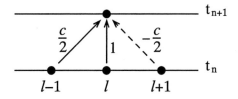

node at time $n + 1$. This means that the right neighbor contributes a negative weight, a sort of *antiparticle* which is the cause of instability.

This is best illustrated by the corresponding spacetime diagram.

How do we cure this problem?

The simplest recipe is to go back to a first order method in space. A popular choice to this end is the *Upwind method.*

3.2.1 *Upwind method*

The upwind method consists in taking a one-sided contribution in the direction the speed comes from (where the wind blows from). So, if $U > 0$, we have:

$$f_l^{n+1} - f_l^n + c[f_l^n - f_{l-1}^n] = 0 \tag{3.10}$$

namely,

$$f_l^{n+1} = A f_{l-1}^n + B f_l^n + C f_{l+1} \tag{3.11}$$

where:

$$A = c, \qquad B = 1 - c, \qquad C = 0, \tag{3.12}$$

The corresponding space-time diagram is given below:

FIG. 3.3. *Spacetime diagram of the upwind method* $(U > 0)$.

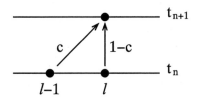

It is readily seen that the positivity condition for the weight reads

$$c < 1, \quad U < U_g \equiv d/h \tag{3.13}$$

namely again the celebrated Courant-Friedrichs-Lewy condition.

Stability is rescued, but to the price of loosing an order of accuracy. In addition, the stencil depends on the sign of the velocity, which is in general changing in space and time: not a likeable situation. We would like to preserve stability without paying this price in accuracy and simplicity. This is achieved by the Lax scheme described next.

3.2.2 Lax scheme

It can be shown that the Euler-Centered scheme for the advection equation corresponds to a time-centered scheme for the following differential equation

$$\partial_t f + U \partial_x f = -\frac{h^2}{2} \partial_{tt} f \qquad (3.14)$$

The reader is kindly encouraged to go through the little algebra that shows the result. The message is quite illuminating: the instability comes from negative diffusion in time as introduced by the spurious anti-Laplacian on the right hand side! Having pinned down the disease, the remedy comes by easily: just balance negative diffusion in time with a corresponding positive diffusion in space!

This is precisely the content of Lax's 'trick':

$$f_l = \frac{1}{2}(f_{l-1} + f_{l+1})$$

namely 'split' the central node into an equal left and right neighbor contributions. With this trick, the centered scheme becomes:

$$f_l^{n+1} - f_l^n + \frac{c}{2}[f_{l+1}^n - f_{l-1}^n] = 0 \qquad (3.15)$$

whose spacetime diagram is given below:

FIG. 3.4. Lax spacetime diagram.

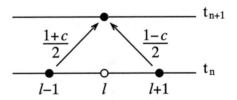

It is easy to show that this scheme is also stable under the CFL condition $c < 1$.

The numerical dispersion relation yields:

$$\Omega = \frac{\arctan[c \tan(kd)]}{h} \qquad (3.16)$$

$$\gamma = -(1 - c^2)(kd)^2 \qquad (3.17)$$

From the latter we see that stability is guaranteed for $c < 1$, but only at the price of introducing artificial smoothing, i.e. *numerical diffusion*. Stability is bought through diffusion and the two requirements are mutually conflicting, a sort of *numerical uncertainty principle*: no pointlike structure can be stable, on the other hand, no extended structure can be diffusion-free!

How to move further on from here?

The recipe is to go to higher order, more complex 'computational molecules', which permit to push discreteness effects from second (diffusion) to higher orders (hyperdiffusion) in kd.

3.2.3 Lax-Wendroff method

A popular choice is the Lax-Wendroff scheme, that we report here for completeness, leaving a detailed analysis to the keen reader.

The Lax-Wendroff scheme is basically a combination of Lax+Leapfrog methods:

Lax:

$$f_{j+1/2}^{n+1/2} = \frac{1}{2}\left[f_j^n + f_{j+1}^n\right] + \frac{c}{2}\left[-f_j^n + f_{j+1}^n\right] \tag{3.18}$$

$$f_{j-1/2}^{n+1/2} = \frac{1}{2}\left[f_j^n + f_{j-1}^n\right] + \frac{c}{2}\left[f_j^n - f_{j-1}^n\right] \tag{3.19}$$

Leapfrog:

$$f_j^{n+1} = c\left[f_{j+1/2}^{n+1/2} - f_{j-1/2}^{n-1/2}\right] \tag{3.20}$$

The resulting space-time diagram = computational molecule is given below

FIG. 3.5. *Lax-Wendroff space time diagram.*

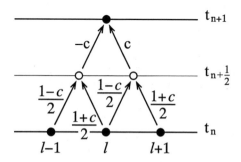

To be noticed the addition of an intermediate time level It can be shown that this achieves zero numerical diffusion and only suffers hyperdiffusion $\gamma h \sim c^2(1 - c^2)(kd)^4$. Of course, this comes at the expense of some complexity.

3.3   Numerical renormalization

All discrete dispersion relations shown so far, save the centered scheme, share a common trait: they reduce to the exact expression $\omega = kU$ in the limit $c = 1$. This means that we don't necessarily need to go to vanishingly small h and d to 'erase' lattice discreteness, these quantities can be both finite provided the fulfill the condition $c = 1$, namely $d/h = U$, which means that signals propagate along the *lightcones* defined by the lattice.

This is not specific of the wave equation, it is a generic principle.

Lattice discreteness effects can be expanded in powers of the 'coupling constant' $\epsilon = c - 1$ and provide a clear identification of artificial numerical effects as expressed by the numerical dispersion relation:

$$\Omega = a_1 k + a_3 k^3 + \cdots + a^{2p+1} k^{2p+1} \tag{3.21}$$

$$\gamma = b_0 + b_2 k^2 + \cdots + b^{2p} k^{2p} \tag{3.22}$$

For the present case, the wave equation, the exact relation is $a_1 = U$, $a_{2p+1} = 0$ and $b_{2p} = 0$. Other terms in the a-series describe *phase errors (numerical dispersion)* and the b-series describe *amplitude errors (numerical diffusion)*.

The 'anomalous' coefficients could obviously be expanded in powers of $c - 1$, and formally analyzed as Feynman diagrams for the interaction of the propagating signal with the lattice. In fact, the scope of higher order schemes is to 'cancel' as many anomalous coefficients as possible. It should be appreciated that the numerical dispersion relation goes beyond such perturbative analysis, since it corresponds to a resummation to *all* orders of anomalous effects. Such a nice property is most naturally formalized via the language of operational calculus.

3.3.1   *Operator identities and "numerical renormalization"*

Differences schemes can be formulated in a very powerful and systematic way by endorsing the language of *operator calculus*. The key operator is the generator of the group of traslations, *shift operator* S_h for brevity:

$$S_h f(x) = f(x + h) \tag{3.23}$$

Equivalent definitions are: $S_h = e^{hd/dx} \equiv (e^{d/dx})^h$

Discrete derivatives can all be expressed as powers of S and its inverse.

- First-order forward:

$$D^+ = \frac{S - 1}{h}$$

- Second-order centered:

$$D^c = \frac{S - S^{-1}}{2h}$$

The relation between Discrete and Continuum dispersion relations takes a closed and very telling expression:

$$D = \frac{\sinh(hD^c)}{h} \qquad (3.24)$$

$$D^c = \frac{\sinh^{-1}(hD)}{h} \qquad (3.25)$$

where $D \equiv d/dx$: Practical use: automatic generation of higher-order schemes by expanding the continuum operator D in powers of the discrete one! Each term of the Taylor expansion is ultraviolet divergent, but the infinite expansion resums to a finite result (renormalization).

Operational expansions can also be given a *telescopic* form:

$$D = \sum_p W_p D_p \qquad (3.26)$$

where $D_p = \frac{S^p - S^{-p}}{2ph}$ is the p-th order centered derivative and W_p are appropriate weights (mass of speed-p particles). For instance, a 4th order finite difference

$$\frac{2}{3h}[f(x+h) - f(x-h)] - \frac{1}{12h}[f(x+2h) - f(x-2h)]$$

can be given a more sensible reorganization in terms of the following 'telescopic' decomposition:

$$D_2 f = D_1 + \frac{1}{3}[D_1 - D_2]$$

To achieve p-th order accuracy all telescopic derivatives up to order p need to be retained in the expansion. These expressions clearly show that higher order accuracy goes to the expense of locality. Again the Heisenberg principle: infinite accuracy requires contributions from the entire energy spectrum.

3.4   Other equations

The concepts illustrated so far are of course not restricted to the case of advection equations. In fact they apply to a huge variety of physical phenomena described by partial differential equations. A few examples are given in the sequel.

3.4.1   *Advection-Diffusion*

The basic principles previously expounded can be used to analyze other physical phenomena, for instance advection-diffusion phenomena:

$$\partial_t f + U \partial_x f = D \partial_{xx} f \qquad (3.27)$$

which plays a crucial role in many applications. Leaving aside convection ($U = 0$) the relevant diffusive Courant number is

$$c_d = Dh/d^2 \tag{3.28}$$

and the stability condition is simply $c < 1/2$. Since the differential operator is second order, centered differences can be safely employed with first-order time marching. In the case of advection-diffusion (i.e. Fokker-Planck) equations, the two Courant numbers add up to further constrain the stability limit

$$\frac{uh}{d} + \frac{Dh}{d^2} < 1 \tag{3.29}$$

From this we see that diffusion is more severe: by doubling spatial resolution, the time step must be reduced by a factor four!

This is why in most applications, advection is treated by explicit method and diffusion by implicit ones.

3.4.2 The Klein-Gordon equation

Another important evolutionary partial differential equation is the Klein-Gordon equation, which describes spinless relativistic bosons:

$$\partial_{tt}\phi - c^2\partial_{xx}\phi = -m^2\phi \tag{3.30}$$

where m is the bosonic mass in atomic units.

It can be shown that centered finite difference schemes produce numerical dispersion in the form of 'slowing-down' effects: the boson wavefunction scatters against the lattice sites and gets consequently slowed down. Non-linear generalizations of the KG equation, such as

$$\phi_{tt} - c^2\phi_{xx} = f[\phi] \tag{3.31}$$

are succesfully handled by centered differences and leapfrog time marching. A simple, and yet very useful scheme for these evolutionary (linear or nonlinear equations) is

$$\phi_l^n - 2\phi_l^n + \phi_l^{n-1} - c^2 dt^2/dx^2[\phi_{l+1}^n - 2\phi_l^n + \phi_{l-1}^n] = f(\phi_l^n) \tag{3.32}$$

with the following space time diagram (note space-time symmetry).

3.4.3 The Dirac equation

The previous anomalies are generally related to breaking of space-time symmetries (the Poincaré group).

Similar considerations apply to more sophisticated internal symmetries such as spin, isospin and gauge symmetries in high-energy physics. For instance, relativistic fermions in 1D are described by the following Dirac equation:

$$\partial_t\psi_i + A_{ij}\partial_x\psi_j = m_{ij}\psi_j \tag{3.33}$$

where i, j are spinorial indices. It can be shown that the numerical dispersion relation of many finite difference schemes as applied to system of equations with internal symmetry produce artificial counterpropagating modes ("fermion-doubling") which represent a subtle problem for the solution of the Dirac equation in a discrete lattice.

FIG. 3.6. *Spacetime diagram for the Dirac equation.*

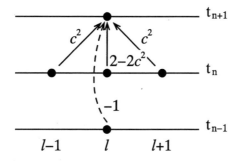

3.5 Boundary Conditions

The previous considerations apply to the interior domain. It is well known that *Boundary Conditions* (BC) play a decisive role in selecting the actual solutions of a given problem because they restrict the class of solutions compatible with the bulk equation (less symmetry).

Three major type of BC's met in the applications are:
- *Periodic*:

$$f(x = 0) = f(x = L) = A$$

- *Dirichlet*:

$$f(x = 0) = A, \quad f(x = L) = B$$

- *Von Neumann*:

$$[af + b\tfrac{df}{dx}](x = 0) = A$$

$$[af + b\tfrac{df}{dx}](x = l) = B$$

where a, b, A, B are generic constants.

From the practical view point, these expressions translate into algebraic equations which must be imposed to the boundary values of the numerical solution.

In one-dimension BC are easy because the boundaries are pointlike. For instance, periodic BC would simply read as follows:

$$f_{L+1} = f_1, \quad f_0 = f_L$$

where $l = 0$ and $l = L+1$ are just computational buffers catering for the missing values at the extreme left and right ends of the lattice.

Neumann boundary conditions require some extra-care. Consider the right-most boundary $x = L$. A second order difference formulation reads:

$$af_L + b\frac{1}{2h}(f_{L+1} - 2f_L + f_{L_1}) = B$$

The problem is that values at f_{L+1} are not defined. The typical recipe is to use a one-sided derivative

$$af_L + b\frac{1}{2h}(f_L - F_{L_1}) = B$$

which is however only first order accurate. This shows that near the boundaries, finite differences may easily loose accuracy. A similar statement applies to non-uniform lattices.

These problems are obviously accrued in higher dimensions, but still manageable as long as the boundaries can be expressed as coordinate lines. This is typically the case for idealized geometries such as cubes, cylinders, spheres and the like. Whenever such a fortunate condition is not met, which is the rule and not the exception in most real-life situtions, the finite-difference method looses much of its power.

Under these conditions, more powerful methods are needed, as we shall illustrate in the next two chapters.

3.6   References

1. T. Pang, An introduction to Computational Physics, Cambridge Univ. Press
2. D. Potter, Computational Physics, Wiley, 1976
3. D. Iserles, Numerical Analsys of differential equations, Cambridge Univ. Press, 1996.

3.7   Projects

1. Solve the Fokker-Planck equation with constant speed U and diffusivity D on a one-dimensional lattice. Examine convergence accuracy as a function of lattice spacing and time-step size for various FD schemes.
2. Same as above, but in dimension D=2.
3. Repeat 1+2 with the Klein-Gordon equation.
4. Repeat 1+2 with the Schroedinger equation.

3.8   Warm-up programs

```
c ==================================
c       advection-diffusion
c ==================================
        parameter (nx=500)
        dimension fold(0:nx+1),fnew(0:nx+1)
c ---------------------------------------------
        write(6,*) 'n.of steps'
        read (5,*) nt
        write(6,*) 'speed and diffusivity'
        read(5,*) u,dif
        write(6,*) 'initial spread'
        read(5,*) delta
        write(6,*) 'centered/Lax: 1,any'
        read(5,*) itype
c initial conditions
        do i=0,nx+1
         x = float(i-1)
         x = (x-nx/2)/delta
         fold(i)=1./(1+x*x*x*x)
         fold(i)=0.
         if(i.gt.nx/2-delta.and.i.lt.nx/2+delta) fold(i)=1.
         write(6,*) i,fold(i)
        end do

        dx=1.
        dt=1.
        c = u*dt/dx
```

```
      d = dif*dt/dx/dx
      write(6,*) 'courant c d',c,d
      pause
      if(itype.eq.1) then
c centered
      al = +0.5*c+d
      am =   1.-2.*d
      ar = -0.5*c+d
      else
c Lax
      al = 0.5*(1.+c)+d
      am = -2.*d
      ar = 0.5*(1.-c)+d
      endif

c time-stepper ------------------------>
      do it=1,nt
c periodic boundary conditions
      fold(0)   =fold(nx)
      fold(nx+1)=fold(1)
      do i=1,nx
       fnew(i)=al*fold(i-1)+am*fold(i)+ar*fold(i+1)
       if(mod(it,10).eq.1) write(7,*) i,fold(i),fnew(i)
c next step
       fold(i)=fnew(i)
      end do
c ------------------------------------->
      end do
      stop
      end
```

4

FINITE VOLUME METHOD

In this lecture we shall expound the basic ideas behind the Finite Volume technique. The main use of this technique is the solution of partial differential equations in complex geometries which cannot be described by global, smooth coordinate systems. This is the case for most real-life geometries, whence the importance of Finite Volumes in practical applications.

FIG. 4.1. *Finite volume grid.*

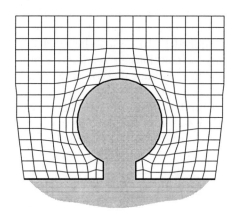

4.1   Finite Volume formulation

The Finite Volume is best formulated for (systems of) equations in conservative form:

$$\partial_t \rho + div \vec{F}(\rho) = S[\rho] \tag{4.1}$$

where ρ is a scalar density, $\vec{F}$ the corresponding flux and S a source term.

The Finite Volume formulation begins by covering the computational domain with a series of congruent trapezoidal cells (Finite Volumes) such that the representation can accurately describe geometrical boundaries of (almost) arbitrary shape.

In local coordinate units, a map taking the original cartesian coordinates (x, y) into a new set of (non-smooth) coordinates (x', y') defined by a series of local mappings such that each trapezoid can be seen as the result of a local deformation of a squarelet of sides dx, dy:

$$x' = a_1 x + a_2 y + a_3 xy \tag{4.2}$$
$$y' = b_1 x + b_2 y + b_3 xy \tag{4.3}$$

By this, the unit square transforms into:

$$(0,0) \rightarrow (0,0)$$
$$(1,0) \rightarrow (a_1, b_1)$$
$$(0,1) \rightarrow (a_2, b_2)$$
$$(1,1) \rightarrow (a_1 + a_2 + a_3, b_1 + b_2 + b_3)$$

Although the new coordinate system (x', y') is no longer smooth, it still is topologically equivalent to a cartesian grid, so that each node or center of the finite volume grid can be uniquely identified and addressed by a pair of integers (i, j). In computational parlance, we say that the finite volume grid is *structured*, namely topologically equivalent to a cartesian grid. This property permits to organize Finite Volume calculations in relatively simple data structures. The great advantage of the Finite Volume method is teh possibility to use body-fitted coordinates, i.e. adapt the mesh to the actual geometry of the problem (beyond cylinders, spheres and similar idealizations!). The mathematical formulation is slightly less intuitive but still very simple. The mathematical core of the Finite Volume method is the Gauss theorem as applied to the conservative equation 4.5:

$$\frac{d}{dt} \int \rho dV + \int \vec{F} \cdot d\vec{A} = \int S dV \tag{4.4}$$

FIG. 4.2. *The finite volume as the local deformation of the unit squarelet.*

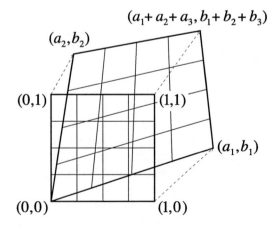

In simple words: *Change rate of the scalar mass in a given volume = Incoming Flux - Outgoing Flux.*

From Gauss' theorem we realize that, at variance with the finite-difference method, the central objects of the Finite-Volume formulation are *volume-averaged* quantities rather than nodal values.

4.2   Discrete Equations of motion

At each centroid C we associate a discrete value ρ_C, $C = 1, \ldots, NV$. Upon applying Gauss theorem to the equation (4.5):

$$\frac{d\rho_C V_C}{dt} + \sum_s \vec{F}_s \cdot \vec{n}_s A_s = S_C V_C \qquad (4.5)$$

where V_C is the cell volume and the index s runs along the contour surface (east-north-west-south boundaries in D=2) of area A_s.

The problem is to evaluate the contour integral, given the fact that discrete density fields are stored in the cell centers. This calls for *Interpolation* from cell surfaces to centers.

With reference to the the two-dimensional grid and its dual depicted in Fig 6.3,

a typical piecewise-linear local interpolator is:

$$F_e = a_{CE} F_C + b_{CE} F_E \qquad (4.6a)$$
$$F_n = a_{CN} F_C + b_{CN} F_N \qquad (4.6b)$$
$$F_n = a_{CW} F_C + b_{CW} F_N \qquad (4.6c)$$
$$F_n = a_{CS} F_C + b_{CS} F_S \qquad (4.6d)$$

where E, N, W, S denote the centers of the East,North,West,South neighboring cells of C. The interpolation coefficients a_{CD}, b_{CD}, $D = E, N, W, S$ depend on the metric of the discrete network. We observe that in general, the face centers

FIG. 4.3. *The finite volume network and its dual represented by cell centers.*

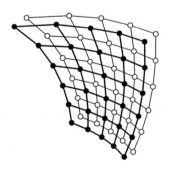

(where the surface integral should be evaluated to obtain 2nd order accuracy) do *not* coincide with the centers of the segments joining C to the centroids of its neighbors. This only happens if the mesh is orthogonal. With this notation, the discrete equation of motion is:

$$\frac{d\rho_C}{dt} + \sum_{D=E,N,W,S} (a_{CD}\vec{F}_C + b_{CD}\vec{F}_D) \cdot \vec{n}_D \frac{A_D}{V_C} = S_C \qquad (4.7)$$

where $\vec{n}_D$ is the normal to the surface crossing the line between centers C and D. This is a set of ordinary differential equations which can be marched in time with a suitable time discretization method. Before discussing time discretization it is worth noting that the above set of equations involves a certain amount of topological information that needs either to be stored or to be computed "on-the-fly" during the computation. We refer to:

Center values: cell volumes V_C
Edge values: projected areas $\vec{n}_D A_D$
Nodal values: grid coordinates $\vec{x}_n$

The same equation also shows that the *collocation* of the different variables leaves some room for choice. For instance, it is a common practice to store scalar quantities at cell centers and vector fields at cell vertices (nodes), so that the interpolation from nodes to edges can proceed using only the values at the nodes which define the segment (surface in D=3) the edge belongs to. For instance, with reference to the east boundary:

$$\vec{F}_e = a_{ne}\vec{F}_{ne} + (1 - a_{ne})\vec{F}_{se}$$

The advantage of this interpolation is that all the three locations involved lie on the same segment. Another possibility is to store vector fields directly at the edge centers, so that they are directly available for the computation of surface fluxes. The schemes in which different variables resides at different locations are called *staggered* and play an important role in the Finite Volume literature. For more details the reader is kindly directed to specialized textbooks on the finite volume method.

4.3   Time discretization

Time can be discretized with any of the methods discussed previously. Once this is done 4.7 becomes a set of algebraic difference equations for the NV unknowns ρ_C. It is easily shown (useful exercise) that for the case of an orthogonal, cartesian mesh, these algebraic equations take the familiar form of 'naive' finite-difference equations.

4.4   Topological constraints: how far can we go?

Finite Volumes provide much more geometrical flexibility as compared to Finite-Differences, but... this flexibility is not unlimited either! In particular, topological 'defects' that must be carefully watched out are:

- Near zero-volume cells:
  Coefficients become singular because of surface-to-volume S/V ratioes going to infinity.
- Overstrechted cells: $dy/dx \to 0, \infty$
  The numerical system becomes numerically unbalanced, with a much stronger coupling along the y direction.
- Kinks:
  Jumps in the derivatives of the coordinate lines, namely, singularities in the curvature of the numerical mesh.
- Warps:
  Non planar volume boundaries, the two triangles defined by nodes A,B,C and B,C,D of a given cell boundary are not coplanar.

All of these defects tend to generate ill-conditioned algebraic problems, i.e the coefficients of the algebraic systems show wild excursions from place to place, so that different regions of the grid must be advanced with different time steps (numerical stiffness).

The generation of a well-balanced *body-fitted* mesh, optimized on the given problem, is a true art, a full-time job (grid-generation software) which takes often more man and computer time than the actual solution of the algebraic equations!

4.5   Computational complexity

Finite-Volume methods are generally more CPU and memory intensive than Finite-Differences because topological information needs to be stored within each volume (cell volumes, surface areas and normal directions and so on...) and processed at run-time. However, in complex geometries this is eclipsed by the significant gain of geometrical flexibility. Finite-Volume is definitely the method of choice in Computational Fluid Dynamics, to solve the Navier-Stokes equations.

$$\partial_t \rho \vec{u} + div[\rho \vec{u}\vec{u} + \mu \nabla \vec{u}] = -\nabla P \qquad (4.8)$$

where $\vec{u}$ is the fluid speed and P the fluid pressure. These are exceedingly difficult equations to solve, due to the infamous problem of Turbulence.

Finite-Volumes is the recommended choice for any problem involving conservative equations in moderately complex geometries. The Finite Volume method is very flexible and still almost as simple as the Finite-Differences method. However it leaves some room to ambiguity: for instance, should vector fields be located at cell centers like scalars, or at face centers so as to facilitate the calculation or boundary fluxes? Many choices are possible, which must be evaluated carefully.

If geometrical complexity is really hard, even more flexible and powerful methods are needed.

This leads us to the subject of the next lecture: Finite Elements.

4.6   Projects

1. Solve the Fokker-Planck equation in two-dimensions using a set of non-uniform rectangles covering a rectangular domain.

2. Same as above with a circular obstacle inside the domain. The solution vanishes on the surface of the circle.

4.7   References

1. M. Vinokur, An analysis of finite differences and finite volume formulation of conservation laws, J. Comp. Phys., 81, p, 1-52, 1989.

2. S. Succi, F. Papetti, An introduction to parallel computational fluid dynamics, Nova Science, NY, 1996.

5

FINITE ELEMENT METHOD

The Finite Element Method (FEM) is the most powerful technique to handle complex geometries and solve partial differential equations in highly complex geometries. Its strengths are geometrical flexibility, as combined with a very solid theoretical basis, rooted into the theory of weak solutions in functional spaces. In this lecture we shall provide the basic elements of the Finite Element method.

5.1 Finite-Element formulation

The starting point of the Finite-Element method is to produce a space *tesselation* using 'unstructured' elements, typically triangles in 2D, tethraedra in 3D, which can flexibly adapt to extremely complex shapes.

The distinguishing topological feature of finite elements is "unstructuredness", namely the coordination number of the mesh can change from place to place like in a amorphous solid. This means that it is no longer possible to address a single node of the FE network with a unique pair/triplet of indices. Instead, the relevant topological information associated with a FE grid is:

- nodes $(d = 0)$
- edges $(d = 1)$
- surfaces $(d = 2)$
- elements $(d = 3)$

As a result, the best way to organize topological data is by numbering node by node, $i = 1, \dots, N$ and associating each node with an *Interaction List* (IL)

FIG. 5.1. *Finite element grid. (oppure) Tessellation of a complex geometrical shape.*

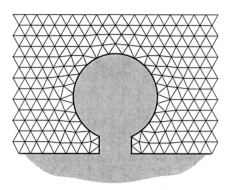

FIG. 5.2. *Topological information associated with a finite element grid.*

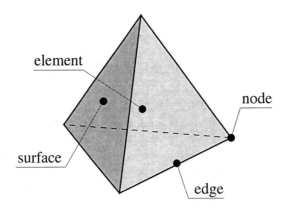

specifying the set of nodes "interacting" with the given node. This information is formalized into the *connectivity matrix*:

$$C_{ij} = 1 \quad \text{if node } i \text{ interacts with node } j \tag{5.1}$$
$$C_{ij} = 0 \quad \text{otherwise} \tag{5.2}$$

The notion of "who interacts with whom" depends on the specific form of the finite-element *basis functions* placed on top of a single node of the FE network. The most natural choice is to define finite element functions which take on the value 1 on a given node and zero at every other node:

$$\psi_i(x,y) = 1, \qquad\qquad x = x_i, \quad y = y_i \tag{5.3}$$
$$\psi_i(x,y) = 0, \qquad\qquad x \neq x_i, \quad y \neq y_i \tag{5.4}$$

In this case the connectivity matrix is non-zero only along nearest-neighbor links. The above expression does *not* imply that finite elements are Dirac's deltas! On the contrary, they are generally *piecewise local polynomials* within each given element. The *support* Ω_i of the finite element $\psi_i(\vec{x})$ is the collection of all elements sharing the node $\vec{x}_i$ as a common vertex. Clearly, the number of such elements is the number of non-zero entries in the corresponding row of the connectivity matrix $z_i = \sum_j C_{ij}$.

As a result, any function $f(\vec{x},t)$ can expanded into the complete set of N finite elements:

$$f(\vec{x},t) \sim f_N(\vec{x},t) = \sum_{i=1}^{N} f_i(t)\psi_i(\vec{x},t) \tag{5.5}$$

with the property $\psi_i(\vec{x}_j) = \delta_{ij}$ and $\psi(\vec{x}) = 0$ for $\vec{x} \notin \Omega_i$.

This is like any other ordinary expansion into a complete set of basis functions, with the crucial property that FE basis functions are *local*, they have finite

FIG. 5.3. *Geometrical support of the finite element basis function.*

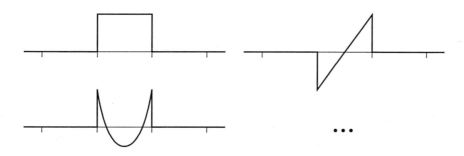

support. As we shall see later, this makes FE's very well suited to numerical computations.

5.2   The equations of motion: projection and weak solutions

The mathematical basis of the Finite-Element method is the theory of weak solutions. The major point is that one no longer requires pointwise convergence, but rather endorse a variational point of view, i.e. require optimality of the solution in a global ("weak" to say it with the mathematicians) sense. In other words, the solution is not required to fulfill the original equation on each node of the discrete lattice: we only require the solution to be an optimal global approximant, in the sense that the distance between the exact solution and the discrete one is minimal in an appropriate functional space. The notion of 'minimal' is best discussed within the framework of of the Galerkin projection method.

Let

$$r_N = [\partial_t - L]f_N \qquad (5.6)$$

be the *residual* associated with the discrete solution f_Ni of the Liouville equation. Strong convergence requires $r_N \to 0$ everywhere as $N \to \infty$. Weak convergence is satisfied with less than this: just require that r_N be minimal within all possible N-dimensional discrete approximants to the exact solution f. It is then intuitive that minimality is achieved if the residual has no components in the N-dimensional space F_N which the solution f_N belongs to. This is tantamount to asking for the following *orthogonality* condition :

$$\langle g|r_N \rangle = 0 \qquad (5.7)$$

where g is any *test* function in the functional space F_N and brackets denote scalar product in F_N. Upon choosing a set of basis functions $\psi_i, i = 1, N$ and test functions $\eta_j, j = 1, N$, Galerkin projection delivers a set of N equations:

$$\langle \eta_j | [\partial_t - L] \psi_i \rangle = 0$$

The residual is orthogonal to the finite dimensional space spanned by the N basis functions $\psi_1 \ldots \psi_N$. By letting test functions run over F_N, that is by covering the geometrical domain with the elements of the Finite-Element method network, a set of N ordinary differential equations is generated:

$$M_{ij} \frac{df_j}{dt} = L_{ij} f_j \tag{5.8}$$

where:

$$M_{ij} = \langle \psi_i | \psi_j \rangle \tag{5.9}$$

$$L_{ij} = \langle \psi_i | L \psi_j \rangle \tag{5.10}$$

where brackets denote scalar product in F_N and basis and test functions have been made the same for simplicity. It is clear that once a given set of basis/test functions is stipulated, any (not necessarily differential) operator receives a specific matrix representation. The nice thing is that finite-support guarantees sparse matrices independently of the order of approximation N. This is in a marked contrast with conventional expansions using global functions, in which the orthogonality constraint forces the basis functions to become highly oscillatory in space, thus undermining numerical stability because orthogonality results from a series of cancellations. Having portrayed the main theoretical ideas behind the Finite-Element method, we now illustrate some simple examples of actual basis functions.

5.3 1D FEM: Hat functions

For simplicity we consider the coverage of a one-dimensional segment $0 < x < L$ with linear finite elements (hat functions).

$$e^-_i(x) = (x - x_{i-1})/(x_i - x_{i-1}), \qquad x_{i-1} < x < x_i \tag{5.11}$$

$$e^+_i(x) = (x_i - x)/(x_{i+1} - x_i), \qquad x_l < x < x_{i+1} \tag{5.12}$$

Note that the support is made by two elements, $e_i = e^-_i + e^+_i$, each being non-zero only within its own support. The lattice has a constant connectivity $z = 2$. We leave as an instructive exercise for the reader to check the following matrix representations for the identity and derivative operators:

$$\text{Identity}: \quad M_{ij} = \frac{h}{6}[1, 4, 1] \tag{5.13}$$

$$\frac{d}{dx} : \quad D_{ij} = \frac{1}{2}[-1, 0, 1] \tag{5.14}$$

$$\frac{d^2}{dx^2} : \quad D_{ij} = \frac{1}{h}[1, -2, 1] \tag{5.15}$$

where we have assumed a uniform lattice $x_i = ih$ for simplicity.

Familiar finite-difference expressions are easily recognized.

The point is that similar expressions hold true also for irregular lattices, without corrupting the optimality (in functional sense) of the Finite-Element method representation! This is to be contrasted with the potential loss of accuracy of finite difference methods on non-uniform grids. It should also be appreciated that the expression (5.9) provides a very neat and systematic way of constructing matrix representations for virtually any operator on fairly general and irregular grids. This should be contrasted with the sort of ambiguities discussed in connection with the Finite Volume method.

5.3.1   *FEM in 2D: The triangle*

The typical unstructured element in D=2 is the triangle. To characterize the basis functions is is convenient to work with local coordinates x, y:

$$x = \frac{X - X_1}{|X_2 - X_1|} \qquad\qquad y = \frac{X - X_1}{|X_3 - X_1|} \qquad (5.16)$$

where capital letters denote absolute coordinates in 2D.

At each vertex we associate a corresponding basis function:

$$\psi_1(x, y) = 1 - x \qquad (5.17)$$
$$\psi_2(x, y) = 1 - y \qquad (5.18)$$
$$\psi_3(x, y) = x + y \qquad (5.19)$$

whose collection around the common vertex generates the pyramidal form typically associated with triangular finite elements.

5.4   Matrix Assembly

An important stage in any Finite Element calculation is the construction of the various matrices associated with the operators. This is called *matrix assembly* and proceeds in two steps:

1. Compute local contributions
2. Assign local contributions to the global matrix elemnt

FIG. 5.4. *Linear finite elements.*

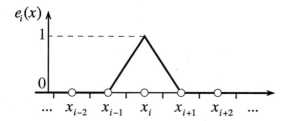

For definiteness, and simplicity, we illustrate these two steps for the simple case of one-dimensional hat elements.

For each element $i = 1, \dots, NE$ compute four local contributions:

Local array:

$$LM_{11} = \langle \psi_i^+ | \psi_i^+ \rangle \tag{5.20a}$$

$$LM_{12} = \langle \psi_i^+ | \psi_{i+1}^- \rangle \tag{5.20b}$$

$$LM_{21} = \langle \psi_i^+ | \psi_{i+1}^- \rangle \tag{5.20c}$$

$$LM_{22} = \langle \psi_{i+1}^- | \psi_{i+1}^- \rangle \tag{5.20d}$$

These integrals can often be computed analytically or, more generally, by low-order numerical quadrature.

Once the contributions are available, they must be mapped into the appropriate global matrix addresses:

$$LM_{11} \rightarrow M_{(i,i)} \tag{5.21a}$$

$$LM_{12} \rightarrow M_{(i,i+1)} \tag{5.21b}$$

$$LM_{21} \rightarrow M_{(i+1,i)} \tag{5.21c}$$

$$LM_{22} \rightarrow M_{(i+1,i+1)} \tag{5.21d}$$

The entire matrix M_{ij} is constructed. by sweeping through the entire lattice. This intimate conection between geometry and functional spaces lies at the heart of the success of the method.

The same procedure, if only a bit more involved, applies to multidimensional situations. In general, an element with p vertices generates a $p \times p$ local matrix LM_e, which is then scattered into the global matrix M_{ij} via an index mapping $IM : e \rightarrow (i,j)$.

FIG. 5.5. *Local and global coordinates in the triangle.*

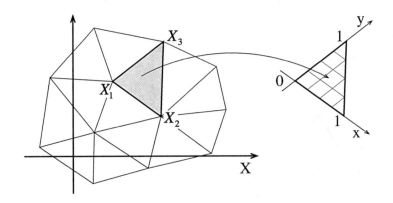

5.5   Boundary conditions

Boundary are geometrically well represented by just adapting the element shape/size to their geometrical form.

Boundary conditions (BC) are typically incorporated into the right hand side of the matrix problem. Von-Neumann boundary conditions are most easily dealt with, since they disappear altogether from the formulation via a mere integration by parts:

$$\langle \psi | div \vec{F}(\psi) \rangle = \langle \nabla \Psi | \vec{F}(\Psi) \rangle = 0 \tag{5.22}$$

if the BC is $\vec{F} = 0$ at the boundary.

These are called "natural boundary conditions".

On the other hand, Dirichlet boundary conditions must be explicitly encoded into the form of the solution. For instance:

$$f(x) = \sum_{n=1}^{N} f_n \Psi_n(x) + \sum_{b=1}^{B} F_b \Psi_b(x_b) \tag{5.23}$$

where n runs over the interior nodes, and b only over the boundary nodes where the Dirichlet condition $f(x_b) = F_b$ is imposed. By taking N scalar products inside the computational domain one generates $B \times N$ contributions from the boundary elements which are then incorporated into the right hand side of the corresponding equation. Dirichlet BC are termed "essential" because they do not disappear upon integration by parts.

Either ways, the procedure handling boundary conditions reveals the same systematic robustness of the Finite-Element method procedure in the interior nodes.

5.6   Computational complexity

The Finite-Element method is normally more expensive than Finite Volumes and Finite Differences because matrix elements have to be computed and then stored into relevant global arrays. Moreover, indirect adressing to keep track of the connectivity is also setting a penalty as compared with simple addressing in cartesian grids. In addition, since the mass matrix is non-diagonal, Finite-Element method generate a matrix problem even for explicit problems!

This is often perceived as a significant disadvantage of the method for highly dynamical problems.

However, the fact that they can accomodate virtually any complex geometry makes of the Finite-Element method the method of choice for heavily complex geometries. Its major field of application is Computational Mechanics.

5.7   FEM for Quantum Field Theory

The traditional area of application for the Finite-Element method is mechanical engineering. Although very slowly, the virtues of the Finite-Element method approach are starting to be appreciated also in different areas. As an aside, we mention that Finite-Element method have been advocated as a useful device for quantum mechanics and quantum field theory calculations. In a nutshell, the idea is to solve the Heisenberg equations for operators (infinite-dimensional matrices in Hilbert space) using Cranck-Nicholson time-stepping. The crucial property is that Cranck-Nicholson time stepping guarantees Equal-Time-Commutation Relations.

$$q_{n+1} - q_n = (p_n + p_{n+1})dt/2 \qquad (5.24)$$
$$p_{n+1} - p_n = (f_n + f_{n+1})dt/2 \qquad (5.25)$$

where p, q are quantum mechanical operators (hence M-dimensional matrices in a finite Fock-space representation). By solving the above equations, it is rapidly shown that $[p_{n+1}, q_{n+1}] = [p_n, q_n] = i\hbar$. By solving algebraically, we obtain explicit expressions of p_n, q_n at time n as a function of initial values p_0, q_0 at time $t = 0$. This implies a lot of symbolic algebra, which currently limits the time span to $n \sim 10$. Alternatively, one can also proceed numerically, in which case the hard-core operation is high-dim matrix-matrix multiplication. The method is intriguing, but it is hard to predict whether it can really compete with more popular methods in statistical mechanics and quantum field theory, such as Path Integration and other Quantum Monte Carlo techniques (see Volume II).

5.8   FEM for quantum gravity: Regge calculus

Finite elements also have a natural connection to Regge calculus in classical and quantum gravity. The idea is to generate dynamic tessellations of the non-Riemannian manifolds associated with general coordinate transformations:

$$x^\mu \to y^\mu = f^\mu[x^\nu] \qquad (5.26)$$

The corresponding metric tensor $g_{\mu,\nu}$ defines the action of the gravitational field as:

$$S_g = \int \sqrt{-g}dx \qquad (5.27)$$

whose minimization delivers the Einstein's equation of general relativity. Dynamic finite elements can be used to discretize the hypersurfaces associated with the metric field and capture the sharp features associated with large distorsions (black holes). Mathematically:

$$g(X, t) = \sum_i g_i(t)\psi_i(X) \qquad (5.28)$$

where X denotes the set of spatial coordinates and tensorial indices have been relaxed for simplicity.

In quantum gravity, many configurations $g_{\mu,\nu}$ are considered in order to construct the path summation over all Feynman stories from a given initial configuration g_I to a final configuration g_F. The probability of each configuration is

$P[g] = e^{-S_g}$ and again, dynamic finite elements can be used to provide a discrete representation of the tensorial field $g_{\mu,\nu}$ well suited to numerical computations.

5.9   References

- G. Strang, G. Fix, An analysis of the Finite Element method, Prentice Hall, Englewood Cliffs, 1973.
- C. Bender, D. Sharp, Solution of operator field equations with the method of finite elements, Phys. Rev. Lett. 50,20, p. 1535, 1983.
- T. Regge, Il Nuovo Cimento, 19, 558, 1961.

5.10   Projects

- Solve the Schroedinger equation in 1D with a localized potential using linear FEM's.
- Same in 2D using triangular elements
- Solve the linear oscillator problem in Heisenberg form using a Cranck-Nicolson FEM formulation.

6

SPECTRAL METHOD

Finite Volumes and Finite Elements address the problem of solving equations in complex geometries. On the other hand, in fundamental science we are often interested in basic mechanisms away from the boundaries. In this case, what we need is top-accuracy in the bulk domain, while boundaries can be handled by periodicity so as to minimize finite-size effects. The best numerical technique to address this type of problems is the Spectral method (SM), which, as suggested by its name, is based on Fourier representation and works therefore in reciprocal (momentum) space. In this lecture we shall cover the basic aspects of this important technique.

6.1 Fourier and Discrete Fourier transforms

The appeal of spectral methods is clear: The Fourier Transform (FT) turns differential operators into simple algebraic polynomials in k-space. Based on the definition of Fourier transform:

$$F[f] \equiv \widehat{f}(k) = \frac{1}{2\pi} \int e^{-ikx} f(x) dx \qquad (6.1)$$

it is well known that:

$$F[df/dx] = -ik\widehat{f}(k).$$

so that derivatives (and all their powers for sufficiently smooth functions) act like simple powers of the wavenumber:

$$(d/dx)^p = (-ik)^p$$

This property turns differential equations into much simpler to solve algebraic equations. The general strategy is then clear:

1. Fourier transform: $\widehat{f} = F[f]$
2. Solve in Fourier space: $\widehat{f} \to \widehat{f}'$
3. Inverse transform back to real space: $f' \leftarrow \widehat{f}'$

where prime means at time $t + dt$. The inverse Fourier transform is defined as:

$$F^{-1}[\widehat{f}] \equiv f(x) = \frac{1}{2\pi} \int e^{+ikx} \widehat{f}(k) dk \qquad (6.2)$$

In order to apply this strategy in actual numerical practice, we first need to define the analogue Fourier (FT) and Inverse Fourier Transforms (IFT) on a discrete lattice.

FIG. 6.1. *Wavenumbers $m = 1$, $m = 3$ and $m = 7$ (which is identical to $m = 1$, because of aliasing) in a 8 points lattice ($N = 8$).*

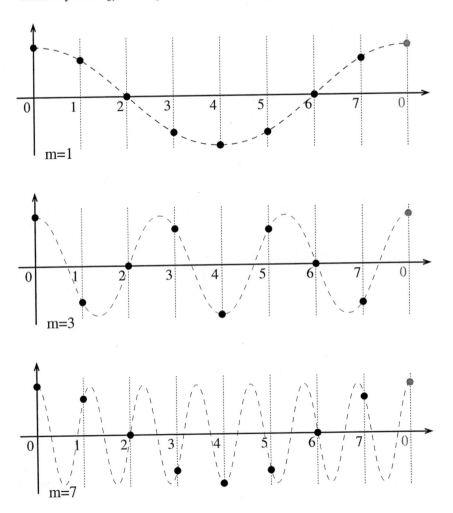

To this end, we introduce a uniform real-space lattice and a corresponding reciprocal lattice in k space:

$$x_l = l dx, \qquad l = 0, \cdots, N-1, \qquad dx = L/(N-1) \qquad (6.3)$$

$$k_m = m dk, \qquad m = 0, \cdots, N-1, \qquad dk = 2\pi/(N-1); \qquad (6.4)$$

Note that in such a lattice the shortest wavelength (highest energy) mode covers two lattice sites.

On these discrete lattices we define the Discrete Fourier Trasform (DFT) as follows:

$$0 \quad 2 \quad 4 \quad 6 \qquad \text{even points}$$

$$1 \quad 3 \quad 5 \quad 7 \qquad \text{odd points}$$

$$\widehat{f}_m = \sum_{l=0}^{N-1} W_N^{lm} f_l \tag{6.5}$$

where

$$W_N = e^{-i2\pi/N}.$$

Which such a representation, differential operators are replaced by (powers of) discrete wavenumbers to *exponential accuracy!* However, this nice property does not come for free. From its very definition, we see that the computation of a N-site DFT requires $O(N^2)$ operations, a very severe computational complexity. This is of course the direct consequence of the *global* nature o the FT: to compute the Fourier transform at any given k *all* values of $f(x)$ are needed! This is the price for turning differential operators into algebraic ones.

Thus, we need a tool to turn around this complexity barrier. This tool, known as Fast Fourier Transform (FFT), exists indeed and represents one of the most important algorithms of applied mathematics ever, with applications across virtually any field of science.

6.2 The magic FFT

To the purpose of definiteness we shall proceed with a practical example. Consider a simple one-dimensional 8-point lattice:

In general:

Even:

$$G_m = \sum_{k=0}^{(N-1)/2} W^{2mk} f_{2k}$$

Odd:

$$H_m = \sum_{k=0}^{(N-1)/2} W^{(2k+1)m} f_{2k+1}$$

The full FT is of course given by the sum of the two:

FIG. 6.2. *The FFT "butterfly"*.

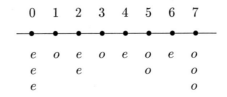

Synthesis:

$$g_m = G_m + H_m$$

These partial transforms exhibit two crucial properties:
Translational Symmetry:

$$G_{m+N/2} = G_m, \ H_{m+N/2} = H_m$$

Scaling Symmetry:

$$W_{N/2}^{2m} = W_N^m$$

These properties open the door to the magic buster of computational complexity:
recursive summation. The idea is to cluster the grid point in recursive classes:

Level 1: Even/Odd (E/O)

Level 2: (EE,EO),(OE,OO)

Level 3: (EEE,EEO),(EOE,EOO),(OEE,OEO),(OOE,OOO)

and so on until we are left with only two elements after $\log_2 N$ stages. The
corresponding tree-structure is given below:

Recursive summation proceeds as follows:

Level 0:

Suppose we focus on a given point, say m: First, we split the FT into an odd
and even components:

$$g_m = G_m + H_m$$

with

$$G_m = W_{m0} f_0 + W_{m2} f_2 + W_{m4} f_4 + W_{m6} f_6$$

and

$$H_m = W_{m1} f_1 + W_{m3} f_3 + W_{m5} f_5 + W_{m7} f_7$$

where $W_{lm} = W_8^{lm}$.

Level 1:

Then we split again ($N' = N/2$, $W' = W_4$):

$$G_m = G_{m0} + G_{m1}$$

with

$$G_{m0} = W_{m0}f_0 + W_{m4}f_4 \equiv W'_{m0}f'_0 + W'_{m2}f'_2$$

and

$$G_{m1} = W_{m2}f_2 + W_{m6}f_6 \equiv W'_{m2}f'_1 + W'_{m2}f'_3$$

Level 2:

A third split delivers ($N'' = N'/2$, $W'' = W_2$):

$$G_{m0} = G_{m00} + G_{m01}$$

with

$$G_{m00} = W_{m0}f_0 \equiv W''_0 f''_0$$

and

$$G_{m01} = W_{m4}f_4 \equiv W''_1 f''_1$$

The G_{m00} is the last leaf of the even branch of the tree, and, by symmetry, H_{m00} is the last leaf of the odd side of the same tree.

Now let's walk the butterfly bottom-up:

At level 3 ($N = 2$), we compute G_{m00} and G_{m01} which contribute to g_m and g_{m+1} by translational symmetry.

At level 2 ($N = 4$), we synthetize $G_{m0} = G_{m00} + G_{m01}$ and compute G_{m1}. By translational symmetry G_{m0} and G_{m1} contribute to g_m and g_{m+2} as well. At this stage 3 of the total 4 partial contributions are accounted for.

At level 1 ($N = 8$), we synthetize $G_m = G_{m0} + G_{m1}$ and by translational symmetry G_m contributes to g_m and g_{m+4} as well. At this stage all four contributions of the even part are accounted for. A completely analogue procedure applies to the odd side of the three. As a result, all we need to complete the computation is to sum up the even and odd sides: $g_m = G_m + H_m$.

It must be noticed that partial contributions can be transferred from one level to the next one without further recomputation *only* because of the scale symmetry of the weights!

This magic algorithm is one of the most important cornerstone in applied mathematics.

6.3   Pseudo-spectral method

The FFT algorithm permits to use the power of spectral methods for *linear, homogeneus* problems with translational symmetry, in which plane waves are natural eigenfunctions.

What about situations, such inhomogeneus or non-linear problems, where homogeneity is broken? Far from being an exception, such situtions are commonplace in most applications, think of the Schroedinger equation with realistic potentials, waves in heterogeneus materials, and many others.

The problem here is that the Fourier transform of the product of the functions is given by the convolution of their Fourier transforms:

$$F[f(x)g(x)] = \sum_{m'=0}^{N-1} \widehat{f}_{m-m'} \widehat{g}'_m \qquad (6.6)$$

Once again, we are confronted with a N^2 complex problem!

This complexity can be beaten by the so called *pseudo-spectral* technique. The basic idea is to first evaluate each function in real space as the Fourier antitransform of their own transform. Then the product is taken in real space and finally Fourier transformed again. In equations:

$$F[fg] = F\{[F^{-1}\widehat{f}][F^{-1}\widehat{g}]\} \qquad (6.7)$$

The obvious advantage is that the N^2-complex convolution is circumvented. However, the procedure introduces numerical errors in the form of *aliases*, namely harmonics which are not representable in the mesh, even though their combination $k' \pm k'' = k$ is.

Mathematically, aliases arise because in discrete Fourier space orthogonality only holds modulo-N:

$$\sum_{lm} e^{i(l+m)\,dk} = \delta(l - m \pm kN), \quad k = 0, 1, 2, ... \qquad (6.8)$$

Aliases can seriously corrupt the solutions by injecting high-frequencies modes (wiggles) which can lead to numerical stabilities.

The recipe to cure them is to replace the plain Fourier transform with the average of shifted transforms by $\pm dx/2$:

$$\widehat{f}_{shift}(k) \frac{F[f(x - dx/2)] + F[f(x + dx/2)]}{2} \qquad (6.9)$$

A direct calculation shows that the two opposite shifts cancel aliases. This is however a rather expensive cure: six extra transforms in $D = 3$. In practice one often simply truncates modes with $|k| > N$. This implies a rather substantial loss of effective resolution, since part of the reciprocal grid is simply wasted, but has the advantage of not requiring any extra-computational effort.

6.3.1 *Operator-splitting methods*

Operator-splitting strategies, working with a double $x - k$ representation are increasingly used in quantum mechanical calculations dealing with standard Hamiltonians of the form:

$$H = -\Delta + V(\vec{x})$$

The kinetic energy operator, Δ, is handled in Fourier space. The Fourier transformed solution is advanced in time, and then transformed back in real space. The real-space solution is then advanced in time again under the action of the potential $V(\vec{x})$ alone.

This *operator-splitting* technique is efficient since it works always in the representation where the relevant partial Hamiltonian is diagonal. Of course, one needs fast switches from one representation to another, but this is not a problem.

6.4 Computational efficiency

The efficiency of Fourier methods is unsurpassed for homogeneus problems in periodic geometries. Perhaps the most typical example is the Poisson equation in a periodic box. Homogeneus incompressible turbulence (Navier-Stokes equations) is also a typical success-case for (pseudo)-spectral methods. Unfortunately, spectral methods do not easily extended to realistic geometries, and also parallel computations (many processors computing simultaneously indepedent portions of the same problem) turn out to be very laborious due to the non-local character of the Fourier transform. Hybrid methods, aimed at combining the resolution of spectral methods with the geometrical flexibility of finite elements, have been proposed in the recent past (Spectral Element method) but only with mixed results.

6.5 References

1. D. Gottlieb, S. Orszag, Numerical analysis of spectral methods, SIAM Philadelphia, 1977
2. S. Orszag, G. Patterson, Numerical simulation of three-dimensional homogeneus isotropic turbulence, Phys. Rev. Lett. 28, p.76, 1972.

6.6 Projects

1. Solve the 1D Schroedinger equation with a space dependent potential $V(x)$.
2. Solve the Burgers equation (see program included) using the Discrete and Fast Fourier trnaform. Compare the computer time as the size of the problem is increased.

6.7 Warm-up program

```
c ssssssssssssssssssssssssssssssssssss
c solves the Burgers equation
```

```
c du/dt + u*du/dx = nu*d^2/dx^2
c sssssssssssssssssssssssssssssssssssss
              parameter (N=64)
              dimension ur (N), ui(N)
              dimension vr (N), vi(N)
              dimension ugr(N),ugi(N)
              dimension vgr(N),vgi(N)
              dimension uur(N),uui(N)
              dimension vvr(N),vvi(N)
c ----------------------------------------
              pi = 4.*atan(1.0)
              u0 = 1.
              vis= 0.01
              dx = 1./float((N-1))
              dk = 2*pi/float(N)
              dt = 0.1*dx/u0
              xl = 0.1
              write(6,*) 'enter n. of timesteps'
              read(5,*) nt
c initial conditions
              iout=0
              do i=1,N
               x = dx*float((i-N/2))/xl
               ur(i)=u0/(1+x*x*x*x)
               ui(i)=0.
               write(60,*) i,ur(i),ui(i)
              end do
c u(k,0)
              call ft(ur,ui,vr,vi,N)
c evolution ----------------------------------------------- >
              do it=1,nt
c FT of grad u
              do j=1,N
               wkj = dk*(j-1)
               vgr(j) =-wkj*vi(j)
               vgi(j) = wkj*vr(j)
              end do
c gradu as antitransforms in real space
              call ift(vgr,vgi,ugr,ugi,N)
c form ugradu
              do i=1,N
               uur(i)=ur(i)*ugr(i)
               uui(i)=ui(i)*ugi(i)
              end do
```

```fortran
c now FT the product
                call ft(uur,uui,vvr,vvi,N)
c and (poor-man) dealias
                Ncut=N
                do j=Ncut,N
                 vvr(j)=0.
                 vvi(j)=0.
                end do
c and finally advance in time
                do j=1,N
                 wkj = dk*(j-1)
                 ome=vis*wkj*wkj
                 T = exp(-ome*dt)
                 vr(j)=vr(j)*(1.-ome*dt)+dt*vvr(j)
                 vi(j)=vi(j)*(1.-ome*dt)+dt*vvi(j)
                end do
                call ift(vr,vi,ur,ui,N)
c output
                if(mod(it,nt/10).eq.1) then
                 iout=iout+1
                 do i=1,N
                  write(60+iout,*) i,ur(i),ui(i)
                 end do
                endif

                write(6,*) 'end time step', it

                end do

                stop
                end
c ==========================================
           subroutine ft(fr,fi,gr,gi,N)
c ==========================================
           dimension fr(N),fi(N),gr(N),gi(N)

           pi=4.0*atan(1.)
           wk= 2.*pi/N
           write(6,*) 'wk,N', wk,N
           f0=1./sqrt(float(N))
           do i=1,N
            gr(i)=0.
            gi(i)=0.
            do j=1,N
```

```
      q=wk*(j-1)*(i-1)
      gr(i)=gr(i)+fr(j)*cos(q)+fi(j)*sin(q)
      gi(i)=gi(i)+fi(j)*cos(q)-fr(j)*sin(q)
     end do
     gr(i)=f0*gr(i)
     gi(i)=f0*gi(i)
    end do

    return
    end
c =======================================
    subroutine ift(gr,gi,fr,fi,N)
c =======================================
    dimension fr(N),fi(N),gr(N),gi(N)

    pi=4.0*atan(1.)
    wk= 2.*pi/N
    f0=1./sqrt(float(N))
    do i=1,N
     fr(i)=0.
     fi(i)=0.
     do j=1,N
      q=wk*(j-1)*(i-1)
      fr(i)=fr(i)+gr(j)*cos(q)-gi(j)*sin(q)
      fi(i)=fi(i)+gi(j)*cos(q)+gr(j)*sin(q)
     end do
     fr(i)=f0*fr(i)
     fi(i)=f0*fi(i)
    end do

    return
    end
c =======================================
      subroutine fft(ar,ai,N,M)
c =======================================
      dimension ar(N),ai(N)
c ---------------------------------------
      pi=4.*atan(1.)
c data in bit-reversed order
      l=1
      do k=1,N-1
       if(k.lt.l) then
        a1=ar(l)
        a2=ai(l)
```

```
      ar(1)=ar(k)
      ai(1)=ai(k)
      ar(k)=a1
      ai(k)=a2
     endif

     j=N/2
     do while(j.lt.l)
      l=l-j
      j=j/2
     end do
     l=l+j
    end do
    write(6,*) 'data bit reversed'
c -------------------------------------------
     l2=1
     do l=1,M
      q=0.
      l1=l2
      l2=2*l1
      do k=1,l1
       u= cos(q)
       v= sin(q)
       q=q+pi/l1
       do j=k,N,l2
        i=j+l1
        a1= ar(i)*u+ai(i)*v
        a2=-ar(i)*v+ai(i)*u
        ar(i)=ar(j)-a1
        ar(j)=ar(j)+a1
        ai(i)=ai(j)-a2
        ai(j)=ai(j)+a2
       end do
      end do
     end do
c normalize
     fnorm=1./sqrt(float(N))
     do i=1,N
      ar(i)=ar(i)*fnorm
      ai(i)=ai(i)*fnorm
     end do

     return
     end
```

```
c =========================================
            subroutine ifft(ar,ai,N,M)
            dimension ar(N),ai(N)
c =========================================
            pi=4.*atan(1.)
c data in bit-reversed order
            l=1
            do k=1,N-1
             if(k.lt.l) then
               a1=ar(l)
               a2=ai(l)
               ar(l)=ar(k)
               ai(l)=ai(k)
               ar(k)=a1
               ai(k)=a2
             endif

             j=N/2
             do while(j.lt.l)
               l=l-j
               j=j/2
             end do
             l=l+j
            end do
c -------------------------------------------
            l2=1
            do l=1,M
             q=0.
             l1=l2
             l2=2*l1
             do k=1,l1
               u= cos(q)
               v= sin(q)
               q=q+pi/l1
               do j=k,N,l2
                 i=j+l1
                 a1= ar(i)*u-ai(i)*v
                 a2=+ar(i)*v+ai(i)*u
                 ar(i)=ar(j)-a1
                 ar(j)=ar(j)+a1
                 ai(i)=ai(j)-a2
                 ai(j)=ai(j)+a2
               end do
             end do
```

```
            end do
c normalize ---------------------------
            fnorm = 1./sqrt(float(N))
            do i=1,N
             ar(i)=ar(i)*fnorm
             ai(i)=ai(i)*fnorm
            end do

            return
            end
```

7

DIRECT METHODS

A large number of problems in science can be cast in the form of simultaneous systems of linear equations. In this lecture we shall deal with procedures to solve these systems by exact (within machine-accuracy) systematic procedures, known as Direct Methods. Direct methods are based on numerical matrix algebra, a classical and huge sector of applied and numerical mathematics. Here we shall only illustrate the prototypical direct method: Gaussian elimination.

7.1 Basic matrix algebra

Let us consider the following set of N linear algebraic equations:

$$\sum_{j=1}^{N} a_{ij}x_j = b_i, \quad i = 1, N \tag{7.1}$$

In symbolic matrix notation, simply:

$$Ax = b \tag{7.2}$$

where A is the $N \times N$ matrix whose elements are a_{ij} and x and b are corresponding N-dimensional arrays. The formal solution of (7.2) is

$$x = A^{-1}b$$

where the inverse matrix elements are given by:

$$a_{ij}^{-1} = \frac{\det A_{ji}}{\det A}$$

where det means determinant and A_{ji} is the cofactor of a_{ij}, namely the $(N-1) \times (N-1)$ matrix obtained from A by dropping the i-th row and j-th column. The above expression formally solves the problem exactly. Unfortunately, it is of virtually no practical use, since the calculation of determinants requires of the order of $N!$ (factorial) operations. Other methods are clearly needed.

The most natural strategy is to proceed by sequential elimination of the unknowns x_1, $x_2 \ldots x_N$ one by one. This elimination strategy is the core of Gaussian elimination methods, which we now describe.

7.2   Gaussian elimination

Gaussian elimination is the paradigm of direct methods. Let us illustrate it with a simple example. Consider a toy 3×3 system:

$$\begin{cases} a_{11}x_1 + a_{12}x_2 + a_{13}x_3 = b_1 \\ a_{21}x_1 + a_{22}x_2 + a_{23}x_3 = b_2 \\ a_{31}x_1 + a_{32}x_2 + a_{33}x_3 = b_3 \end{cases} \tag{7.3}$$

In the following we assume that the system is non-singular, namely

$$\det A \neq 0$$

To solve by successive elimination, we first start by eliminating the first unknown x_1:

$$\begin{cases} E_1 \\ E_1 \dfrac{a_{21}}{a_{11}} - E_2 \\ E_1 \dfrac{a_{31}}{a_{11}} - E_3 \end{cases} \tag{7.4}$$

where E_1, E_2, E_3 is a shorthand for the three equations above.

This leaves us with a new matrix A'

$$\begin{cases} a'_{11}x_1 + a'_{12}x_2 + a'_{13}x_3 = b'_1 \\ a'_{22}x_2 + a'_{23}x_3 = b'_2 \\ a'_{32}x_2 + a'_{33}x_3 = b'_3 \end{cases} \tag{7.5}$$

with zeroes along the first column except the a'_{11} element. By applying a similar transformation to the new system, we end up with an upper triangular matrix A''

$$\begin{cases} a''_{11}x_1 + a''_{12}x_2 + a''_{13}x_3 = b''_1 \\ a''_{22}x_2 + a''_{23}x_3 = b''_2 \\ a''_{33}x_3 = b''_3 \end{cases} \tag{7.6}$$

Such a system can be easily solved algebraically proceeding in reverse order from the bottom equation upwards. The above process can be formalized in terms of a sequence of matrix transformations:

$$A \rightarrow A' \rightarrow A'' \dots A^{(N)}$$

The transformation between two successive elements of this sequence is given by:

$$a'^{(n)}_{ij} = \sum_{k=i+1}^{N} \frac{a_{ik}^{(n-1)} a_{kj}^{(n-1)}}{a_{ii}^{(n-1)}} \tag{7.7}$$

After N such transformations, we end up with a upper triangular matrix which is easily solved with a reverse sweep. If, occasionally, a diagonal element is found to be zero, a search along the row is performed until a non-zero element, a_{ij}, is found. A permutation between j-th and i-th columns resets the procedure, a technique called *pivoting*. Having assumed that the matrix A is non-singular, we are guaranteed that a non-zero diagonal element a_{ii} can always be found. This procedure is exact and, up to round off, and it guarantees that the solution is reached after N elimination steps.

The procedure is not a fast one, though. Since each transformation requires $O(N^2)$ matrix-matrix multiplications, the Gauss elimination algorithm is $O(N^3)$ in complexity.

Another reason that makes Gaussian Elimination unpractical for large-scale calculations is memory occupation.

To better appreciate this point, let us focus our attention on *sparse* matrices, namely matrices in which a large majority of elements are zero. More precisely, let ν represent the maximum number of non-zero elements per row:

$$\nu = \max_i \left[\sum_j |sign[a_{ij}]| \right]$$

For a full matrix, $\nu \sim N$, for a sparse matrix $\nu \ll N$.

Typically, full matrices encode global constraints, for instance integral equations with long-range kernels, whereas sparse matrices are typically associated with the discretization of partial differential equations.

Next, we introduce the notion of *bandwith B*, namely the maximal distance in logical space between two non-zero matrix elements.

$$B = \max_{\substack{i,j; \\ a_{ij} \neq 0}} |j - i|$$

It is important to appreciate that in multi-dimensional applications B is generally *much larger* than ν.

In fact, that since computer memory is topologically linear, there is no way all physically contiguous unknowns can be stored into logically contiguous memory locations. The result is that the bandwidth is strongly sensitive to the way unknowns are numbered in a multidimensional domain. The example below illustrates the idea.

Let us consider a *sparse* matrix arising from the discretization of a multidimensional differential operator, say a 2D Poisson problem for simplicity:

$$[\partial_{xx} + \partial_{yy}]f(x,y) = g(x,y) \tag{7.8}$$

Simple centered differences generate a 5-point stencil ($dx = dy = 1$):

$$
\begin{array}{ccc}
 & 1 & \\
1 & -4 & 1 \\
 & 1 &
\end{array}
$$

with only 5 non-zero elements per row. Consider now a 10×3 set of unknowns numbered in lexicographic xy order as follows:

FIG. 7.1. *XY typewriter ordering of a 2D set of 30 unknowns. The bandwidth associated with a simple nearest-neighbor scheme is $B = 10$.*

$$21\ 22\ 23\ 24\ 25\ 26\ 27\ 28\ 29\ 30$$
$$11\ 12\ 13\ 14\ 15\ 16\ 17\ 18\ 19\ 20$$
$$1\ \ 2\ \ 3\ \ 4\ \ 5\ \ 6\ \ 7\ \ 8\ \ 9\ 10$$

It is clear that the maximum *logical* distance between two non-zero elements is 10, a typical matrix row looking like this:

$$\ldots \text{zeroes}\ d\ 0\ 0\ 0\ 0\ 0\ 0\ 0\ l\ m\ r\ 0\ 0\ 0\ 0\ 0\ 0\ 0\ 0\ u\ \text{zeroes} \ldots$$

$$d, l, r, u: \text{ resp. down, left, right, up neighbors;}$$

$$m: \text{ mid, actual unknown;}$$

The same problem with yx lexicographic order gives a much more compact matrix structure, with bandwidth $B = 3$

FIG. 7.2. *YX typewriter ordering of a 2D set of 30 unknowns. The bandwidth associated with a simple nearest-neighbor scheme is $B = 3$.*

$$3\ 6\ 9\ 12\ 15\ 18\ 21\ 24\ 27\ 30$$
$$2\ 5\ 8\ 11\ 14\ 17\ 20\ 23\ 26\ 29$$
$$1\ 4\ 7\ 10\ 13\ 16\ 19\ 22\ 25\ 28$$

with the following matrix pattern:

$$\ldots \text{zeroes}\ l\ 0\ 0\ d\ m\ u\ 0\ 0\ r\ \text{zeroes} \ldots$$

$$d, l, r, u: \text{ resp. down, left, right, up neighbors;}$$

$$m: \text{ mid, actual unknown;}$$

For a banded matrix of bandwidth B, CPU complexity is $O(NB^2)$, and memory occupation is $O(NB)$.

Clearly, we would like to store only non-zero elements. The problem is that the *non-linear* matrix transformation (7.7) generates non zero-values within the bandwidth region also where the original matrix had simply zeroes, the so-called 'fill-in" problem. Fill-in explains the sharp focus on optimal numbering schemes minimizing the bandwidth in the direct solution of linear system. For complex domains this is a NP complete problem, in the same class of the famous traveling salesman problem: find the shortest path visiting a set of N cities.

7.2.1 *The LU factorization*

Gaussian elimination falls within the broader class of *matrix factorization* algorithms for the solution of linear systems. The scope of a factorization algorithm is to represent matrix A as the product of auxiliary matrices $A_1, A_2, \ldots, A_n$ such that each of them can be solved algebraically. The simplest and most popular instances are LU and LDU factorizations, where L, D and U are lower triangular, diagonal and upper triangular matrices respectively.

It is easy to show that Gaussian Elimination performs the LU decomposition. In fact, Gauss algorithm can be cast in the form of a series of $N - 1$ matrix products:

$$A_1 = L_1 A, \qquad A_2 = L_2 A_1 \qquad \cdots \qquad A_{N-1} = L_{N-1} \ldots L_1 A \qquad (7.9)$$

where L_n are lower triangular (left) matrices, so that their product is also lower-triangular. By writing this product formally as L^{-1} we obtain $A = LU$, q.e.d.

Once the LU decomposition is found, the solution is readily found as follows: Let $y = Ux$, then solve

$$Ly = b \qquad \text{(direct sweep)}$$

for the unknown y and:

$$Ux = y \qquad \text{(reverse sweep)}$$

to obtain the sought solution x.

7.3 Gaussian elimination in 1D: the Thomas method

For the case of one-dimensional problems with nearest-neighbor connections, Gaussian elimination reduces to a simple explicit algorithm known as the Thomas method. The problem is to solve the following set of equations:

$$a_l x_{l-1} + b_l x_l + c_l x_{l+1} = g_l \qquad (7.10)$$

This involves tridiagonal matrices of the form:

$$\begin{pmatrix} b_1 & c_1 & & & & \\ a_2 & b_2 & c_2 & & & \\ & a_3 & b_3 & c_3 & & \\ & & \ddots & \ddots & \ddots & \\ & & & & c_{N-1} \\ & & & & a_N & b_N \end{pmatrix}$$

Such matrix problems are typically generated by one-dimensional Sturm-Liouville equations:

$$\frac{d}{dx}[a(x)y' + b(x)y] + c(x)y = g(x) \tag{7.11}$$

$$y(0) = A, \qquad y(L) = B \tag{7.12}$$

The Thomas algorithm starts by postulating a solution in the form:

$$y_{l+1} = A_l y_l + B_l \tag{7.13}$$

By plugging this into (7.10) we obtain:

$$(b_l + A_l c_l)y_l + a_l y_{l-1} = b_l - B_l c_l \tag{7.14}$$

which provides the identities:

$$A_{l-1} = \frac{-a_l}{b_l + A_l c_l} \tag{7.15}$$

$$B_{l-1} = \frac{g_l - B_l c_l}{b_l + A_l c_l} \tag{7.16}$$

Upon specification of the boundary conditions, for instance simple Dirichlet: $y_N = B, y_0 = A$, one proceeds from A_N, B_N (available from B) to A_{N-1}, B_{N-1} down the line up to A_1, B_1 (backward sweep). With all coefficients A_l, B_l available an upward sweep starting from A (from boundary conditions) all the way up to B, using the recursion (7.13) delivers all the unknowns.

The Thomas algorithm is important not much as a self-standing solver of 1D problems, but rather because it is often part of reduction strategies for multi-dimensional problems. For instance, 3D problems of size N^3 can often be handled as an iterative sequence of N^2 one-dimensional problems of size N.

7.4   Software considerations

Although a general understanding of basic direct algorithms is important, it is nonetheless *not* recommended to code up these algorithms on our own. Linear algebra solvers problems are best handled by standardized, professional software offering better performance and reliability. Nowadays, several good packages are available for free at www sites, see for instance: *netlib.ornl.gov*.

7.5   References

I. Duff, A. Erisman, J. Reid, Direct methods for sparse matrices, Oxford Univ. Press, 1989.

7.6   Projects

- Solve the 1D Poisson problem with Dirichlet Boundary Conditions using the Gaussian elimination procedure.
- Compare performances with a time-dependent solution using Fourier methods
- Same as above in D=2

8

ITERATIVE METHODS

Direct methods to solve linear systems of algebraic equations face with an extremely steep complexity barrier in computational complexity as the size of these systems is increased. The result is that, already in three-dimensional applications, direct methods become prohibitively expensive. The most effective way out to this problem is provided by iterative procedures. These procedures are based on the idea of looking no longer for the exact solution, but for approximate solutions to a given level of accuracy. By adjusting this level of accuracy, as well as the starting conditions, iterative procedures can often come to the desired solution much faster than direct methods. In this lecture we shall discuss some major classes of iterative methods.

8.1 Generalities

Iterative methods are based on the idea of looking for *approximate* solutions, sometimes called *quasi-solutions*, to a given linear algebraic problem $Ax = b$. A quasi-solution x_ϵ is defined by the inequality:

$$|Ax_\epsilon - b| < \epsilon \tag{8.1}$$

where ϵ is the prescribed tolerance.

The major idea behind iterative methods is that clever heuristics can get to the desired quasi-solution much faster than direct methods.

It is very convenient to view iterative procedures as explicit maps in a fictitious *iterative* time:

$$x^{k+1} = Tx^k \tag{8.2}$$

where T is the transfer operator mapping iteration level k into the next one, $k + 1$. Starting with a first guess, x^0, we seek an optimal trajectory getting within the stipulated neighborhood of the solution in the least possible number of iterations. The notion of neighborhood is specified by the distance between the actual iterate x^k and the exact solution x^*:

$$d(x^*, x^k) < \varepsilon \tag{8.3}$$

where $d(\cdot, \cdot)$ denotes some suitable metric in the functional space F_N of the approximate solution. Superscript 'star' labels the exact solution and ε is the stipulated tolerance.

FIG. 8.1. *Iterative trajectory getting in the prescribed neighborhood of the exact solution.*

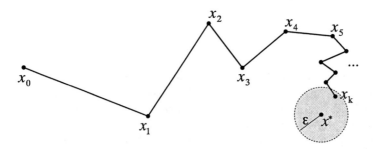

8.2   Theoretical Conditions for convergence

The basic question with iterative methods is of course: *under which conditions does a given iterative method converge?* The theoretical answer is fairly intuitive: like for explicit methods, the iterative map must be stable, i.e. errors in the initial guess must decay (as fast as possible).

The corresponding mathcmatical requirement is:

$$||T|| < 1 \qquad (8.4)$$

where $|| \cdot ||$ is the norm of the operator, namely $||T|| \equiv Max_y\{|y = Tx|\}$, $|x| = 1$. Such a condition is readily found by requesting the absolute error E^k (see next section) to decay in time. Fast error decay is obviously equivalent to fast convergence.

In a finite-dimensional space of dimension N the transfer operator is represented by a $N \times N$ matrix T_{ij}, which gives rise to a quadratic form:

$$\langle x, Tx \rangle \equiv \sum_{ij=1}^{N} T_{ij}x_i x_j = \text{const.} \qquad (8.5)$$

We shall consider *positive-definite* matrices, obeying the constraint:

$$\langle x, Tx \rangle \geq 0 \qquad (8.6)$$

for which contour levels are closed, elliptic curves.

Convergence is strictly related to the *spectral properties* of the transfer operator. In the first place, as a necessary condition, the real part of the eigenvalues of the matrix $-T$ must be negative.

A basic property which heavily affects the convergence rate of iterative methods is the *Condition number* defined as the ratio between maximum and minimum eigenvalues:

$$\kappa[T] = |\lambda_{max}/\lambda_{min}|. \tag{8.7}$$

To appreciate the role of the condition number it is useful to start from a spectral representation of the matrix T:

$$T = \sum_m P^m \lambda_m \tag{8.8}$$

where P^m is the projector upon the m-th eigenvector of T.

From this definition, it is clear that the component of a generic vector x along V^M, the eigenvector associated with the largest eigenvalue, is magnified/contracted by a factor λ_M. The eigenvectors of the matrix give the principal axes directions, and therefore the condition number is a measure of the 'anisotropy' of the contour levels. The ideal condition is $\kappa = 1$ which corresponds to circular contours. High condition numbers are associated with *sensitivity to round-off errors*, namely with 'needle-shaped' contour levels derivatives are very strong and generate a very undesirable sensitivity to small errors/inaccuracies. We shall return to these geometrical considerations when discussing gradient methods.

8.3   Convergence criteria

Several metric notions are useful to put the notion of convergence on quantitative grounds:

- *Absolute Error:* $E^k = d(x^k, x^*)$;
- *Relative Error:* $e^k = E^k/|x^k|$;
- *Relative Deviation:* $d^k = d(x^{k+1}, x^k)/|x^k|$;
- *Residual:* $r^k = Ax^k - b$.

In principle, the relative error is the most reliable theoretical estimate of convergence. The problem is that since the exact solution is not known, the relative error can*not* be measured in actual practice! This is why it is often replaced with the relative deviation, which is perfectly measurable at any given iteration cycle. On the other hand, the relative deviation makes reliance on the assumption that small relative departure means small relative error. This is certainly plausible reasonable, but not foolproof! In fact, such an assumption breaks down in the presence of metastable states where the system does not move much from an iteration to the other even though it is still far away from the exact solution.

This is where the notion of residual becomes useful. By plugging the accepted solution x^k into the original system $r^k = Ax^k - b$ we can estimate how far we really are from solving the problem exactly (to machine round off). Therefore, it is always a safe practice to evaluate the residual before accepting the quasi-solution. If the residual is not small enough, the solution can be iteratively refined as follows (*Richardson's refinement*):

$$A\delta x = -r^k \qquad\qquad (8.9)$$

$$x^{k+1} = x^k + \delta x \qquad\qquad (8.10)$$

Incidentally, Richardson's refinement is a commendable practice also for direct methods!

Having clarified these basic notions, we move on to the description of a few major iterative algorithms. We shall discuss the following categories:

- *Operator splitting methods*
- *Relaxation methods*
- *Gradient methods*

8.4   Operator Splitting methods

Operator splitting methods are based on suitable decompositions of the matrix A. Let $A = P + Q$ be such a binary decomposition, a typical iteration reads as:

$$Px^{k+1} = -Qx^k + b \qquad\qquad (8.11)$$

where we have kept the right hand side one cycle behind.

What do we gain from such reformulation?

Essentially the fact that matrix P might be much easier to invert/solve than the original matrix A! This is a mathematical criterion. Physically, the matrix P would normally be associated to *tight coupled* variables, whereas Q would associate with *loosely coupled* variables. For instance, a diffusion problem with much stronger diffusivity along, say x, directions would yield the following matrix:

$$
\begin{array}{ccc}
 & D_y & \\
D_x & (1 - 2D_x - 2D_y) & D_x \\
 & D_y &
\end{array}
$$

If $D_x \gg D_y$, it makes sense to identify matrix P with the horizontal links and matrix Q with the vertical ones. By doing so, the 2D matrix problem is reduced to a sequence of loosely coupled 1D problems in which the tightly-coupled unknowns are solved exactly via, say, the Thomas algorithm.

Let us now examine some classical examples of splitting schemes.

8.4.1   *Jacobi splitting*

Jacobi splitting, the simplest one, corresponds to the following decomposition:

$$P = D,$$
$$Q = -(L + R).$$

In other words, all non-diagonal links are treated as 'weak' and kept one cycle behind in the fake-time iterative evolution. The result is unsurpassed simplicity since the algebraic problem is now trivially diagonal!

$$a_{ii}x_i^{k+1} = -\sum_{j\neq i} a_{ij}x_j^k + b_i \qquad (8.12)$$

The convergence of this scheme is good only for strongly *diagonal-dominant* matrices, namely

$$|a_{ii}| > \sum_{j\neq i} |a_{ij}| \qquad (8.13)$$

Such a condition requires "good" operators, such as Laplacian, and, more generally, fast-decaying interactions.

8.4.2 Gauss-Seidel splitting

Jacobi iterations are very simple but not very efficient. An improved version, due to Gauss-Seidel, consists in the following splitting:

$$P = L + D,$$
$$Q = -R.$$

In other words, data available from the previous iteration are immediately used to solve for the next iteration:

$$\sum_{j>i} a_{ij}x_j^{k+1} + a_{ii}x_i^{k+1} = -\sum_{j<i} a_{ij}x_j^k + b_i \qquad (8.14)$$

This scheme is more tightly coupled than Jacobi, and it can be shown to converge twice faster. Still, to provide satisfactory results it also requires nice and well-behaved matrices.

8.4.3 Successive Under and Over-Relaxation

Both Jacobi and Gauss Seidel have a rather slow decay rate. In fact, it can be shown that their decay rate goes to zero for high-frequency components of the error field (this observation lies at the heart of the celebrated multigrid methods).

Significant speed-up is obtained by the under/over-relaxation technique.

The idea is to keep a balanced weight between levels k and $k+1$:

$$x^{k+1} = (1-\omega)x^k + \omega x^{k+1} \qquad (8.15)$$

where $0 < \omega < 2$ is the *relaxation parameter*.

Distinguished situations are:

- $0 < \omega < 1$: SUR (Successive Under-Relaxation)
- $1 < \omega < 2$: SOR (Successive Over-Relaxation)

SUR corresponds to a 'conservative' strategy, giving more weight to the past iteration value and it is therefore useful in the presence of oscillatory behaviour to convergence. SOR corresponds to 'aggressive' behaviour, more weight to the future, and it is therefore suitable for slow monotonic convergence.

FIG. 8.2. *Iterative trajectory getting in the prescribed neighborhood of the exact solu-tion.*

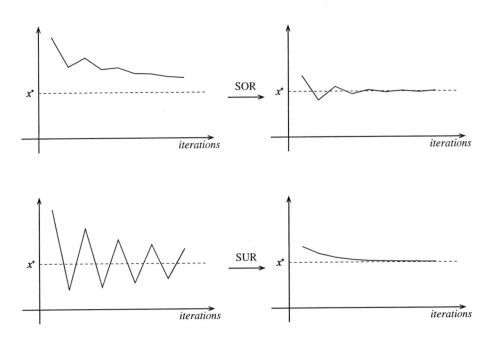

Careful fine-tuning of the relaxation parameter can lead to very significant gains in convergence speed. Unfortunately, the optimal value ω_{opt} is unknown beforehand for all but the simplest matrices. In addition, speed-up is very peaked around the optimal value ω_{opt}, in the sense that even small departures from this optimal value can severely degrade the performance. Therefore SOR techniques, very popular in the 70'-80's, and still useful today, have nonetheless been super-seded by more powerful minimization techniques, to be discussed in the sequel.

8.4.4   Tsebitchev acceleration

An interesting variant of SOR and SUR is to make allowance for a cycle-dependent relaxation parameter ω_k. Typically the k dependence is adjusted in such a way as to maximize the rate of decay of the residual. This can bring dramatic savings, but, again, for general matrices it is very hard to pin down the correct relaxation schedule beforehand.

8.4.5   Block-SOR/SUR iterations

SOR and SUR techniques as expounded so far work at the level of the single nodal unknown x_i. More powerful versions dealing with *cumulative blocks* of unknowns, say all variables along a grid line, can yield a significant boost in performance. Let us refer to *block tri-diagonal* matrices such as the ones generated by, say, a

FIG. 8.3. *A single block* $A_j = D_j + T_j + U_j$. *L_j corresponds to link with 'down' variables in row $j - 1$ and R_j to 'up' variables at $j + 1$.*

$$
\begin{pmatrix}
d & & & & m\ r & & & u & & \\
& d & & & & l\ m\ r & & & u & \\
\cdots & & d & \cdots & & & l\ m\ r & \cdots & & u & \cdots \\
& & & d & & & & l\ m\ r & & & u \\
& & & & d & & & & l\ m & & u
\end{pmatrix}
$$

Poisson problem in 2D:

It is clear that all lmr unknowns in a given j-th row can be treated simultaneously, while unknowns at $j \pm 1$-th rows can be brought to the right hand side. This is the so-called *Block-Jacobi* procedure.

In matrix notation:

For all $j = 1, N_y$:

$$P_j = T_j \qquad \text{(the } lmr \text{ left-middle-right block)}$$
$$Q_j = -(D_j + U_j)$$

where T_j is the *block-tridiagonal* matrix (to be solved by Thomas algorithm) associated with unknowns on the j-th row of the grid. Block-Gauss-Seidel schemes can be applied with the very same procedure, yielding significant savings over their corresponding point-wise counterparts.

8.5 Dynamic relaxation methods

We already observed that iterative methods can be viewed as explicit methods in fake-time, the idea being that fictitious trajectories can take the system to convergence faster than the actual physical evolution. In a way, we can say that while physical trajectories minimize the action $S = \int L dt$, fictitious trajectories must minimize the computational time to steady state. In other words, we can think of a fictitious evolutionary problem in the form

$$\dot{x} = -Ax + b \equiv -r \tag{8.16}$$

whence it is clear that at steady state $\dot{x} = 0$ the original algebraic problem $Ax = b$ is automatically satisfied. This simple observation spawns a whole array of fake-time *relaxation* schemes in the form:

$$x^{k+1} - x^k = \tau_k(b - Ax^k) \tag{8.17}$$

where τ_k plays the role of a relaxation parameter. The corresponding transfer operator is:

$$T_\tau = I - \tau A \tag{8.18}$$

where index k has been omitted for simplicity. It is clear that these are explicit methods in disguise. Consequently, they are subject to similar restrictions in the size of the time-step. Nonetheless, a clever guess on the initial solution can make them much faster than direct methods. It is left as an exercise to the reader to prove that splitting-relaxation methods described previosuly can be recast in relaxation form. The idea of fictitious time evolution gives its best in connection with *dynamic minimization* methods, as described in the following.

8.6   Gradient methods

Let us go back to the idea of the solution $Ax = b$ as the steady state of a fictitious dynamical system evolving according to

$$\dot{x} = b - Ax \tag{8.19}$$

The steady state of this dynamic system can be associated to the minimum of the following quadratic functional

$$Q(x, x) = \frac{1}{2}(x, Ax) - (x, b)$$

In fact, $\nabla Q = Ax - b$, so that the solution of $Ax = b$ is an extremum of Q. If the matrix A is positive-definite such an extremum is also a global minimum. It is therefore sensible to look for a geometric path, or trajectory, in the N-dimensional vector space V_N, getting into a suitable neighborhood of this minimum in the least possible number of steps. A very intuitive strategy is to proceed along the direction where Q decreases most, the so-called *steepest descent*. By definition, this is given by the gradient of Q, which also coincides with the *residual* of the solution:

$$r_k \equiv \nabla_k Q = Ax^k - b \tag{8.20}$$

The typical steepest-descent move looks then like follows:

$$x^{k+1} = x^k - \alpha_k \cdot r_k \tag{8.21}$$

The question is: how to determine the scalars α_k?

Simply on the condition that the restriction of $Q(x, x)$ to the segment pointing along r^k out of x^k be minimum at x^{k+1}, namely:

$$\frac{dQ[\alpha^k]}{d\alpha} = 0 \tag{8.22}$$

Simple algebra yields:

$$\alpha_k = \frac{\langle r_k, r_k \rangle}{\langle r_k, Ar_k \rangle} \tag{8.23}$$

where matrix A has been assumed symmetric positive-definite (SPD). The relations (8.21+8.23) complete the steepest-descent algorithm.

A few considerations are in order.

The SD scheme can be seen as a relaxation process taking the system to global equilibrium under the drive of the following dissipative (Lyapunov) functional:

$$\Lambda(x) = \frac{1}{2}\langle x, x \rangle \qquad (8.24)$$

In fact, the dynamical system 8.19 delivers

$$\frac{d\Lambda}{dt} = -Q(x, x) \qquad (8.25)$$

or, equivalently,

$$d\Lambda = -Qdt = -\nabla Q \frac{dx}{dt} dt = -\langle r, r \rangle dt$$

This expression clearly shows that the SD dynamics is guaranteed to lower the Lyapunov functional: a genuine H-theorem!

This is a very nice property. However, a number of cautionary remarks are in order.

First, the SD algorithm is guaranteed to converge only for SPD matrices. Any deviation from such conditions may send the trajectory away from the solution rather than getting closer to it. It is also clear that, like any other iterative scheme, the SD algorithm is sensitive to the condition number of matrix A. Small round-off errors can generate local runaway moves which slow down or may even disrupt convergence altogether.

Second, from expression (8.6) it is clear that only the earliest SD iterations are very fast since Λ is "big" far from equilibrium, but they become increasingly slower as the exact solution x^* is approached.

Therefore, although the underlying physical picture of a dissipative process erasing errors according to a genuine H-theorem is very appealing, all in all, SD *cannot* be regarded as very efficient iterative solvers. As we shall see, slight generalizations of the SD philosophy prove significantly faster.

8.7 Conjugate-Gradient methods

Let us rewrite the basic equation $Ax = b$ in the following form

$$Ax' = 0 \qquad (8.26)$$

where $x' \equiv x - x^*$ is the 'non-equilibrium' departure from the the exact solution, x^*, which fulfills $Ax^* = b$ by default.

Removing primes for simplicity, we now consider a point x_0 lying on the level curve (ellipse) $(x_0, Ax_0) = C_0$. We would like to get to the exact solution $x = 0$, the bottom of the Q-valley, corresponding to $C = 0$, with a *single move*. This is generally *not* the case if we decide to follow the steepest descent direction. It becomes possible if we move along the *conjugate gradient* direction instead.

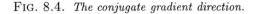

FIG. 8.4. *The conjugate gradient direction.*

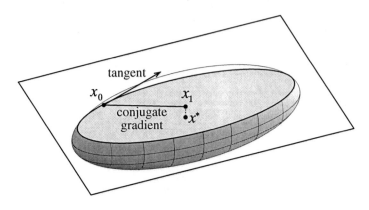

The conjugate-gradient direction y at point x_0 is defined by the so-called A-orthogonality condition:

$$(x_0, Ay) = 0 \tag{8.27}$$

Geometrically, y is the vector aligned with the tangent to the ellipsoid $(x, Ax) = C_0$ at point x_0.

The tangent is orthogonal to the gradient and A-orthogonal to the conjugate gradient, the latter being the direction joining x_0 to the origin $x = 0$. The strategy is therefore to choose the search direction between two successive iterations not along the steepest descent but along the conjugate descent instead. This guarantees that on each plane $[x^k, x^{k+1}]$ the local minimum yields $C_{k+1} = 0$. By choosing a sequence of linearly independent search directions we are therefore guaranteed to reach the exact solution in *at most* N steps.

The CG algorithm proceeds as follows:

1. Initialization: choose initial guess and first search direction

$$x_0, p_0 = r_0 \equiv Ax_0 - b \tag{8.28}$$

2. Next search direction is taken as a combination of the previous one and the actual residual

$$p_1 = r_1 + \beta_1 p_0 \tag{8.29}$$

The parameter beta is then fixed by A-orthogonality between p_0 and p_1.

$$\langle p_1, Ap_0 \rangle = 0 \tag{8.30}$$

which delivers

$$\beta_1 = -\frac{\langle r_1, p_1 \rangle}{\langle p_0, Ap_0 \rangle} \tag{8.31}$$

This condition secures linear independence because the so-called *Krylov* sequence $\{p_0, Ap_0, A^2 p_0 .... A^k p_0\}$ is shown to be linearly independent and therefore it spans a vector space $K_N[A]$ known as Krylov space.

3. Once beta is known we proceed to the next iterate according to the usual procedure

$$x_1 = x_0 + \alpha_1 p_1 \tag{8.32}$$

with α given by the usual expression:

$$\alpha_1 = \frac{\langle r_1, p_1 \rangle}{\langle p_1, Ap_1 \rangle} \tag{8.33}$$

Iteration of steps 1+2+3 takes to the exact solution in at most N steps (modulo roundoff errors of course). Often, it goes much faster if the first guess is chosen in a "close" subspace of x^*. The CG algorithm is very powerful and widely used in many fields of modern computational physics wherever large systems of equations need to be solved. Recently, it has found wide use also in eigenvalue problems generated by electronic structure calculations. Among others, it has the formal advantage of being applicable also to complex geometries since it does not assume any regular ordering of the unknowns (like for instance in block Gauss-Seidel).

Of course, it is not failsafe either. In particular, like the SD method, it is supposed to work well only for SPD matrices, such as those associtaed with elliptic problems (diffusion processes). For mixed problems (advection-diffusion) more sophisticated variants, such as *Bi-conjugate* or *Generalized Residual Methods* are required. These variants are based on the idea of using multiple search directions simultaneously, so that moves along the "wrong" direction can be corrected "on the fly". A very promising modern direction relates to *genetic algorithms* in which many simultaneous search directions are activated at any given step, and subsequently promoted/suppressed according to some "fitness" criteria based on a cost function, typically the Lyapunov functional. The reader fond of more details is kindly directed to the ponderous literature on the subject

8.8   Preconditioning

We have mentioned several times that the convergence of iterative methods is significantly affected by the Condition Number

$$\kappa[A] = |\lambda_{max}/\lambda_{min}|$$

The Perfect Matrix has a condition 1 and corresponds to (a multiple of) the identity: its contour levels are circles and a *single* iteration takes any initial guess into the exact solution!

It is therefore clear that matrices with near-one condition should provide much faster convergence. This consideration lies at the heart of the so-called *preconditioning* techniques. The idea of preconditioning is to trasform the original matrix A into a new matrix $A' = PA$, where P is the preconditioner matrix, such that $\kappa[A']$ is much closer to 1 than $\kappa[A]$.

$$PAx = Pb, \quad \longrightarrow \quad A'x' = b' \qquad (8.34)$$

Ideally $A' = I$, but it is clear that achieving this goal means solving the linear system! Therefore, one generally looks for semi-empirical approximates. The simplest preconditioner is just the diagonal part of A (Jacobi preconditioning):

$$P_{ij} = diag(A) \equiv a_{ii}{}^{-1}\delta_{ij}$$

More powerful preconditioners are based on Incomplete LU decomposition:

$$PA = I + E \qquad (8.35)$$

where E is an error matrix With $P = (LU)^{-1}$ we would achieve $A' = I$. Therefore, one generally starts factorizing the matrix A (a very expensive task, since it basically like solving $Ax = b$ with a direct method!), but stops at some intermediate step without completing the procedure, whence the diction "Incomplete". Clearly, preconditioned iterators are a mix of direct and iterative methods. Again, this is a huge chapter of numerical analysis to which the reader is directed for in depth details. Here we content ourselves with signalling the idea that preconditioning is often a very powerful strategy to accelerate convergence of iterative solvers.

8.9   Multigrid methods

We conclude this chapter on iterative methods with a brief discussion of a very powerful technique known as *Multigrid*. Multigrid methods build upon the observation that the rate of error decay is a sensitive function of their *wavelength(frequency)*. In other words, short wavelengths are more effectively damped than long ones (think of the dissipative Laplace operator Δ whose effect goes quadratically with the wavenumber k). In fact, long wavelengths are the main culprit for slow convergence on high resolution grids. It makes therefore sense to devise a strategy involving *Multiple Grids* at various resolution and move around different levels so as to remove fast modes on coarse grids and refine the calculation on the fine ones. Let us be a bit more specific.

8.9.1   *Multigrid formulation*

Consider the standard problem $Ax = b$ and imagine to solve it with two-level procedure involving a coarse and fine grid respectively. On the fine grid we have:

$$A_f x_f = b_f \qquad (8.36)$$

Suppose we perform a certain number n_f of Gauss Seidel iterations, leaving us with a residual

$$r_f = A_f x_f - b_f \tag{8.37}$$

This residual is affected by long-wave components which die hard on the fine grid, so that it makes little sense to insist further with fine-grid iterations. A smarter strategy is to 'transfer' the residual to the coarse grid, relax it over there, and then move back the new residual, free of long-wave components (which are short on the coarse grid!). To this end, we need a pair of twin operators moving information from/to fine to/from coarse levels.

8.9.2 Fine-to-Coarse: Restriction

The FTC (Fine-to-Coarse) transfer is called *Restriction*. Suppose for simplicity the coarse grid is just twice less dense than the fine one, the restriction operator is just halving the components of any vector it acts upon. The restriction operator, R maps a N-component vector f in the fine grid onto a $N/2$-component vector c in the coarse grid.

$$c = Rf \tag{8.38}$$

In matrix representation R is a rectangular $N \times N/2$ matrix. The simplest form of restriction, called "injection", consists of assigning to the coarse grid just every other value in the fine grid:

$$c_i = f_{2i-1}, \qquad i = 1, \dots, N/2 \tag{8.39}$$

where c, f denote a generic vector in the coarse/fine grid respectively.

FIG. 8.5. *A 9-point fine vector is restricted to a 5-point coarse vector.*

fine vector: $(\times \; \times \; \times \; \times \; \times \; \times \; \times \; \times \; \times)$
$\qquad\qquad\quad \downarrow \quad\; \downarrow \quad\; \downarrow \quad\; \downarrow \quad\; \downarrow$
coarse vector: $(\times \quad \times \quad \times \quad \times \quad \times)$

More convenient forms of restriction involve some weighting, for instance:

$$c_i = \frac{f_{2i-1} + f_{2i}}{2}, \; i = 1, N/2 \tag{8.40}$$

8.9.3 Coarse-to-Fine: Extension

The reciprocal operation, coarse to fine, is of course an extension of vector c into f, typically achieved via an interpolation operator P:

$$f = Ec \tag{8.41}$$

A simple interpolation may look like follows:

FIG. 8.6. *A 10-point fine vector is restricted to a 5-point coarse vector.*

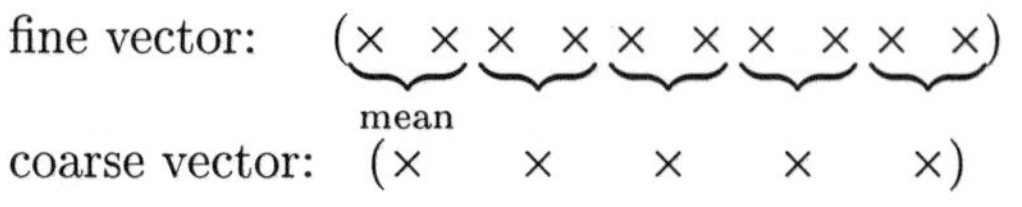

fine vector: $(\times \;\; \times \times \;\; \times \times \;\; \times \times \;\; \times \times \;\; \times)$

mean

coarse vector: $(\times \qquad \times \qquad \times \qquad \times \qquad \times)$

FIG. 8.7. *A 5-point coarse vector is extended to a 10-point fine vector.*

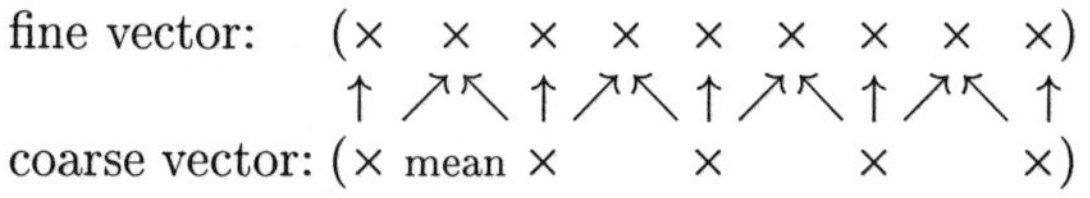

fine vector: $(\times \;\; \times \;\; \times \;\; \times \;\; \times \;\; \times \;\; \times \;\; \times \;\; \times)$

$\uparrow \; \nearrow\!\!\!\nwarrow \uparrow \; \nearrow\!\!\!\nwarrow \uparrow \; \nearrow\!\!\!\nwarrow \uparrow \; \nearrow\!\!\!\nwarrow \uparrow$

coarse vector: $(\times \;\; \text{mean} \times \qquad \times \qquad \times \qquad \times)$

$$f_{2i} = \frac{c_i + c_{i+1}}{2}, \qquad i = 1, \dots, N/2 \tag{8.42}$$

Clearly, E is represented by a $\frac{N}{2} \times N$ rectangular matrix. A desirable property is that R and P are exact 'pseudoinverses', namely: $R^{-1}E = I_N$ and $E^{-1}R = I_{N/2}$, where I denotes the identity.

8.9.4   *Multigrid implementation*

We are now in a position to specify the practical steps of a basic multigrid procedure.

Let:

$$r_f = A_f x_f - b_f \tag{8.43}$$

be the residual in the fine grid, which we aim to reduce by iteration. First, we move it to the coarse grid by means of the restriction operator

$$r_c = R r_f \tag{8.44}$$

Then we solve for the correction x_c in the coarse grid:

$$A_c x_c = -r_c \tag{8.45}$$

Next we move the correction back to the fine grid:

$$y_f = E x_c \tag{8.46}$$

and update the fine grid solution:

$$x'_f = x_f + y_f \tag{8.47}$$

Now we need a new residual r'_f, hopefully much smaller than r_f and, above all, free of nasty long wavelengths. By mere definition:

$$r'_f = A_f x'_f - b_f \tag{8.48}$$

and using the previous definitions

$$r'_f = A_f(x_f + y_f) - b_f = r_f + A_f y_f = r_f + E x_c \tag{8.49}$$

This leaves us with

$$r'_f = r_f - A_f E A_c^{-1} r_c \tag{8.50}$$

which is:

$$r'_f \equiv F r_f = (I - A_f E A_c^{-1} R) r_f \tag{8.51}$$

This is the desired expression of the operator filtering out the long-wave components of the residual.

Having discussed the various steps of the procedure, the last thing we need to specify is the multigrid schedule, i.e. the journey across the various multigrid levels.

8.9.5 *Multigrid schedule*

The typical schedule consists of a series of iterations at various levels of increasing coarseness and then a return leg to the original fine grid, a typical so-called "V-cycle". To be noted that at the bottom of the cycle, the problem should be small enough to be solved "easily", typically by a direct method, or even analytically!

FIG. 8.8. *A V-cycle with 4 levels: one fine, three coarse.*

```
fine level     ×                      ×
coarse lev.         ×            ×
coarser                  ×    ×
still coarser            ×
                    ─────────────────→
                      iterations
```

Convergence is only tested at the two extremal 'f' points in the cycle. If satisfactory, the procedure stops, if not, another V-cycle is performed yielding a so-called W-cycle.

It should be appreciated that a single iteration on a coarse grid at level l in D dimensions, ($l = 0$ being the fine grid), is a factor 2^{lD} cheaper than the same iteration on the fine grid. This *exponential* factor says it all on the enormous potential of multigrid methods as a down-cutter of computational

FIG. 8.9. *A W-cycle made of 2 V-cycles with 4 levels: one fine, three coarse.*

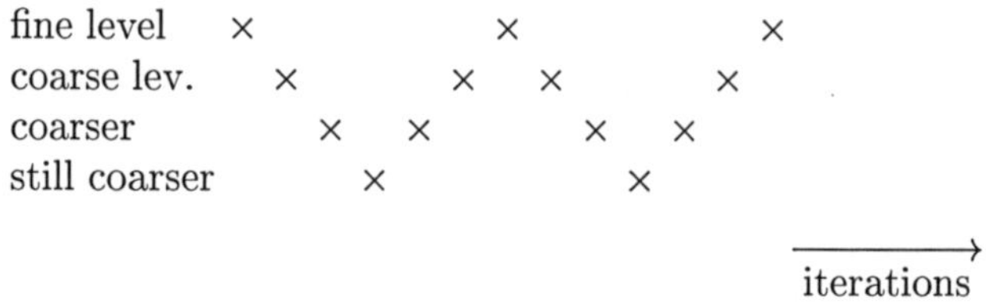

```
fine level    ×              ×              ×
coarse lev.     ×         ×   ×         ×
coarser           ×   ×         ×   ×
still coarser       ×             ×
                                                    ⟶
                                              iterations
```

costs! Multigrid methods represent a very powerful and general tool in modern applied mathematics, with many applications in physics and engineering, ranging from the solution of complex three-dimensional partial differential equations, to the calculations of the electronic structure of large molecular systems.

8.10   Iterative or Direct?

We have already stressed many times in these lectures that iterative methods are often the only viable option for very large systems. Their major advantages are:

- Memory: Only non-zero elements need to be stored;
- Time: By a clever choice of the initial guess and the "right" tolerance, they can get to the right solution on a time-span which scales only linearly, and often even sub-linearly, with the size of the problem;

However:

- Convergence: Convergence is not guaranteed especially for poorly conditioned matrices;
- Robustness: convergence may be highly sensitive to round-off errors;

As a thumb rule, in one-dimension direct methods are generally the best choice, in two-dimensions there's matter of debate, and from three-dimensions upwards, iterative methods are the only viable option.

8.11   References

- A. Iserles, Numerical analysis of differential equations, Cambridge Univ. Press, 1996
- O. Axelsson, Iterative Solution methods, Cambridge Univ. Press, 1994
- W. Briggs, A multigrid tutorial, SIAM Philadelphia, 1987.

8.12   Projects

- Solve the Poisson problem in 2D using as many iterative methods as you can.
- Compare the performances as a function of grid size.
- Same with a convection-diffusion equation.

9

SYSTEMS OF NON LINEAR EQUATIONS

Many problems in science and engineering involve the solution of a set of simultaneous non-linear equations. In this lecture we shall present a cursory view of the major techniques to solve generic sets of non-linear algebraic equations.

9.1 General problem

Let us consider a set of simultaneous non-linear equations of the form:

$$F_i(x) = 0, \quad i = 1, N \tag{9.1}$$

where $x \equiv [x_1, \ldots x_N]$ is a N-dimensional state vector in, say, R^N. Any solution of the above system is also called a *zero-point* of the *map* $F_i(x)$.

A non linear map can generally exhibit *many* zero points, which are called *attractors* or *repellors* depending on whether they are locally *stable* or *unstable*. Stable means that, as a result of a local perturbation around the zero-point, the perturbed trajectory goes asymptotically back to its zero-point. The opposite is true for an unstable zero-point.

Linear systems are clearly a special case of 9.1 with $F_i = \sum_j a_{ij} x_j - b_i$.

We shall discuss the following main classes of methods:

1. *Fixed-point methods*
2. *Gradient methods*
3. *Dynamic relaxation methods*
4. *Genetic methods*

9.2 Fixed-point methods

The idea of fixed-point methods is to define an iteration map whose fixed-point is precisely the solution x^* of the equations $F(x) = 0$ (index i suppressed for simplicity).

Define:

$$G(x) = x - F(x) \tag{9.2}$$

The solution of 9.1 is equivalent to finding the fixed-point of $G(x)$ (the zero-point of F is the fixed-point of G). In geometrical terms, the problem (9.2) is to find the intersection between the two curves $y = x$ and $y = G(x)$. This naturally suggests the following iterative procedure *(Picard iterator)*:

$$x^{(k+1)} = G(x^k) \qquad k = 0, 1, \cdots, k_{max} \tag{9.3}$$

The geometrical picture corresponding to Picard iterations is given below:

From this picture we see that the Picard iteration only converges if the map G is locally contractive, namely:

$$\frac{|G(x') - G(x)|}{|x' - x|} < C < 1$$

where C is some positive constant less than 1. To understand this condition it is sufficient to write down the change between two successive iterates:

$$x^{k+1} - x^k = \frac{G(x^k) - G(x^{k-1})}{x^k - x^{k-1}}(x^k - x^{k-1})$$

and require that the error should decay in the course of the iteration. If the map is (locally) non-contractive the "trajectory" runs away from the fixed point.

The prototype of ideal function for fixed-point iteration is the square-root. Let, $F(x) = \frac{1}{2}(A/x - x)$, which is equivalent to $x^2 = A$ for any non-zero x. This yields $G(x) = \frac{1}{2}(x + A/x)$ and it a useful exercise to verify that Picard iteration is an excellent algorithm for the numerical computation of $\sqrt{A}$.

Fixed-point iterations are generally faster as the exact solution is approached. As a result, they are particularly useful whenever high accuracy is desired.

9.3 Newton-Raphson method

The Newton (or gradient) method is based on a Taylor expansion in a local neighborhood of an initial guess x^0, hopefully close enough to the exact solution x^*.

$$F_i(x) = F_i(x^0) + J_{ij}^0(x_j - x_j^0) \tag{9.4}$$

FIG. 9.1. *Geometrical picture of Picard iterations.*

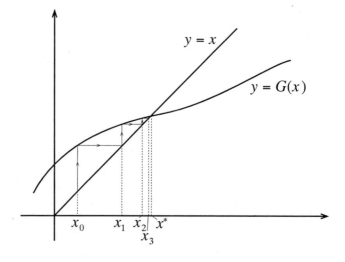

where:

$$J_{ij}^0 = \frac{\partial F_i}{\partial F_j}\bigg|_{x^0} \tag{9.5}$$

is the Jacobian matrix evaluated at the guessed solution point. The resulting Newton iterator is therefore:

$$x^{(k+1)} = x^k - J^{-1}F(x^k) \qquad k = 0, 1, \cdots, k_{max} \tag{9.6}$$

where subscripts have been suppressed for simplicity. The geometrical picture of Newton iteration is given in figure 9.3.

FIG. 9.2. *Geometrical picture of Newton iterations.*

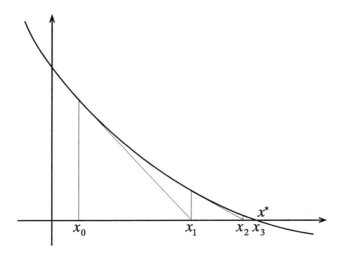

From this picture, as well as from the equation 9.6 we see that Newton iterations may fail to converge if the Jacobian matrix becomes nearly singular (local minimum).

A useful recipe to obviate this problem is to expand to second order:

$$F_i(x) = F_i(x^0) + J_{ij}y_j + H_{ijk}y_jy_k \tag{9.7}$$

where we have set $x = x^0 + y$ and H_{ijk} is the Hessian matrix formed with second order derivatives. This expansion defines a corresponding second-order effective Jacobi matrix:

$$J_{ij}' = J_{ij} + \sum_k H_{ijk}y_k \tag{9.8}$$

which is (most likely) no longer singular. Of course, this is more expensive than a normal Jacobi step. Convergence of the Newton method is generally fast (when

it works!). In addition, the method extends naturally to multidimensional problems. Of course, accurate and efficient ways of computing the Jacobian matrix
J_{ij} are key to success of this method.

9.4 Dynamic relaxation methods

We have repeatedly observed that iterative methods are coneveniently viewed
as dynamic trajectories in fictitious time. The type of dynamics discussed so far
work well if the landscape associated with the "force" F_i is sufficiently smooth,
i.e. if F_i are smooth functions of their argument. Many situations in complex system dynamics involve so-called "corrugated landscapes" in which the function F
undergoes rapid and localized variations, with many interweaved attractors and
repellors. Under such conditions, it is convenient to switch to more physically-
oriented type of dynamics, both *deterministic* and *stochastic*. Either ways, the
starting point is to identify F_i as a N-dimensional force vector, so that the exact solution of the nonlinear systems corresponds to the zero-force points of a
corresponding dynamics.

9.4.1 *Deterministic dynamics*

Simply interpret the equations (9.1) as the steady-state of a classical Newtonian
system:

$$m \frac{d^2 x_i}{dt^2} = F_i(x) \tag{9.9}$$

where we shall assume for simplicity that the "force" F_i derives from a potential.
This is certainly the case for many important problems in physics, such as finding
the ground-state of quantum many-body systems, where the issue is to find the
minimum of the energy functional. Let us therefore assume that F_i derives from
a gradient function, which we identify with the potential energy of the N-body
system. For such a system we define a conserved energy as the sum of kinetic
plus potential energies:

$$E = E_K + E_P = \frac{mv^2}{2} - V(x) \tag{9.10}$$

where $v^2 = \sum_i (dx_i/dt)^2$. We recognize that the solution of the non linear system
9.1 corresponds to the minimum of the potential energy $F_i = -\partial_{x_i} V = 0$. Since
total energy is conserved, the minimum of the potential energy corresponds to the
maximum of the kinetic energy. Therefore, we can explore the 'phase-space' of the
system by 'launching' an ensemble of walkers-particles moving around according
to the above Newton equations. By monitoring the time evolution of kinetic and
potential energy for each walker as a function of time it is therefore possible to
locate the global minima/maxima, or at least a reasonable neighborhood. Once
this local neighborhood is identified, a "conventional" Newton iterator can be
conveniently used to refine the solution.

Of course this is *not* a cheap way of locating minima, and much wisdom is required to keep the number of walkers at a minimum by exploring only "important" regions of phase space.

Keeping the ensemble of walkers to a reasonable size runs the unavoidable risk of missing the interesting regions of phase-space because many trajectories might get trapped into local minima. This is why deterministic Newtonian mechanics is generally not very effective as a dynamic relaxation method.

It is often more effective to move to stochastic methods.

9.4.2 *Stochastic (Langevin) dynamics*

The main idea of stochastic methods is provide the system with the capability of escaping local minima. This is generally realized via "up-hill" moves which locally increase the potential energy but open up new landscapes to the dynamic trajectory. The most typical stochastic mover is based on the Langevin equation, which is based on two ingredients:

1. Accelerate attraction to fixed-points by addition of a dissipative force,
2. Counteract local trapping by addition of a stochastic, noisy force allowing the system to move up-hill along the potential energy landscape.

The result is the following Langevin dynamics:

$$m\frac{d^2x}{dt^2} = F(x) - \eta v + \widetilde{F} \tag{9.11}$$

where $\widetilde{F}$ is a stochastic noise with zero mean and a tunable variance, basically the temperature T of the heat bath in a contact with the system. The viscous term is responsible for kinetic energy decrease, so that the system is more rapidly attracted to minimum points. On the other hand, if these attractors are unwanted local minima, the noise term $\widetilde{F}$ gives them a chance to escape and restart new trajectories hopefully heading towards better attractors. It is therefore clear that the variance of the noise force plays the role of a control parameter, the temperature, which permits to fine-tune the escape rate from local minima. More sophisticated strategies, such as *Simulated Annealing* are based on clever dynamic schedules $T = T(t)$ designed in such a way as to reach the optimum even in the presence of very rough landscapes.

9.5   Genetic methods

The dynamic relaxation methods discussed above perform a sort of Lagrangean mapping of the energy landscape by spreading around 'swarms' of walkers and let them evolve according to Newtonian or Newton-Langevin dynamics.

A kind of Eulerian analogue is represented by genetic search algorithms. The basic idea is as follows.

Lay down a random distribution of points $\{x_p\}$, $p = 1, NP$ and evaluate the corresponding values of the potential $V_p = V(x_p)$. The points scoring the lowest values of V_p are more likely to be close to the global minimum, whereas those scoring high V_p are presumably far from it. It makes therefore sense to put a penalty on "poor-performers" and promote the "good ones". In other words, the population with high V_p is depressed/suppressed whereas "good" points spawn, say $n_p > 1$ "children", located in a neighborhood of their parent's location x_p. As the iteration proceeds, generations of sample points are produced, which are likely to cluster around the local minimum in a naturally adaptive mechanism. This type of adaptive search is best suited to highly localized (golf-hole) potentials.

9.6 Projects

- Compute the value of $\sqrt{5}$ using fixed-point iteration
- Same with Newton-Raphson algorithm
- Find the minimum of the 'soliton-like' potential $V(x; h) = -\cosh^2[\frac{x-a}{h}]$ using the methods described in this chapter. Which one is the best?

9.7 References

- R. Fletcher, Practical methods of optimization, Wiley, London, 1987.

10

EIGENVALUE PROBLEMS

Eigenvalues and eigenvectors describe the natural modes of dynamical systems, like the deformation of an elastic solid, or the oscillations of a quantum wave-function. Therefore, they are paramount across virtually all problems in science and engineering. In this lecture we shall cover some major techniques to locate eigenvalues and eigenvectors of a generic matrix.

10.1   General problem

Eigenvalue problems arise from integral or differential equations with associated boundary conditions. A prototypical eigenvalue problem is generated by the Sturm-Liouville system:

$$Ly \equiv \frac{d}{dx}[y' + p(x)y(x)] + q(x)y(x) = \lambda y(x) \quad y(0) = A, \quad y(1) = B \quad (10.1)$$

Once translated into a numerical language, the corresponding matrix eigenvalue problem reads as follows:

$$Ay = \lambda y \qquad (10.2)$$

where A is the $N \times N$ matrix associated with the Sturm-Liouville operator L, y is the N-dimensional state-vector and λ the sought set of eigenvalues.

Here N denotes the number of discrete degrees of freedom, typically the number of grid-points. We shall assume that boundary conditions, crucial to the determination of the eigenvalues, are automatically incorporated in the matrix A.

The spectrum of the continuum problem generally contains a countable infinity of eigenvalues, whereas the corresponding discrete spectrum consists of at most N discrete eigenvalues. It is clear that the missing eigenvalues are associated with natural eigenmodes which oscillate too rapidly to be representable in the discrete grid. Most often, the task of the numerical eigenvalue problem is not to find the entire discrete spectrum,

$$\sigma_N = [\lambda_1 \ldots \lambda_N]$$

but only a much smaller set of lower-lying eigenvalues

$$\sigma_n = [\lambda_1 \ldots \lambda_n]$$

with $n \ll N$. This constitutes a great simplification, and it permits to handle very huge problems with millions or billions of degrees of freedom, which would otherwise be far beyond reach of even most powerful supercomputers.

Of course, we should make sure that these n numerical eigenvalues properly converge to their continuum value once the mesh size is sento to zero.

10.2　Mathematical problem

Given a $N \times N$ matrix A, we shall make abstraction of the physical origin of such matrix and focus on the mathematical eigenvalue problem of finding a set of scalars λ and N-dimensional vectors x such that:

$$Ax = \lambda x \tag{10.3}$$

In other words, the action of matrix A on a given eigenvector is simply to multiply it by a factor λ, its corresponding eigenvalue, with no rotation. The solution of (10.3) delivers the *spectrum*:

$$\sigma_N = [\lambda_1, \ldots \lambda_N] \tag{10.4}$$

and the corresponding set of eigenvectors

$$E_N = [e_1, \ldots e_N] \tag{10.5}$$

The algebraic condition to solve (10.3) is:

$$p_N[\lambda] \equiv \det[A - \lambda I] = 0 \tag{10.6}$$

This shows that the eigenvalues are the roots of the *characteristic polynomial* $p_N(\lambda)$. These roots are generally complex numbers, with the important exception of Hermitian operators ($a_{ij} = a_{ji}^*$) for which they can be shown to be real.

In the simple instance of a discrete countable spectrum, the characteristic polynomial can be written as:

$$p(\lambda) = \prod_{k=1}^{M} (\lambda - \lambda_k)^{m_k} \tag{10.7}$$

denoting $M < N$ distinct roots of multiplicity $m_1, \ldots, m_M$, $\sum_{k=1}^{M} m_k = N$.

The corresponding set of eigenvectors spans the discrete space $R^N = Span[e_1, \ldots e_N]$

An eigenvalue of multiplicity m generates m generalized eigenvectors spanning an m-dimensional *invariant subspace* I_m of dimension m. The action of A on any vector x in this invariant subspace leaves the vector within the same subspace, but with x and Ax no longer aligned.

In order to compute, say, eigenvector e_k we have to solve the linear system $Ae_k = \lambda_k e_k$, $k = 1, M$. Therefore, finding the entire spectrum is clearly a very heavy-duty, basically $O(N^3)$ complex, computational task.

As per the eigenvalues, based on (10.6) one could think of simply locating the roots of the characteristic polynomial with some suitable root-finder as discussed in the previous chapter. This only works in practice if N is sufficiently small, say $N \ll 10$. For much larger N (easily of the order of 10^6 in structural engineering and $N = 10^{12}$ in electronic structure calculations) other methods must be found. We shall discuss three possibilities:

1. *Direct methods*
2. *(Inverse) iteration methods*
3. *Dynamic minimization*

10.3   Direct eigenvalue solvers: Housolder rotations

The idea of direct eigenvalue solvers is to find a transformation T taking the matrix A into a diagonal form:

$$\tilde{A} = T^{-1}AT = \Lambda \qquad (10.8)$$

For instance, with $N = 3$:

$$\Lambda = \begin{pmatrix} \lambda_1 & 0 & 0 \\ 0 & \lambda_2 & 0 \\ 0 & 0 & \lambda_3 \end{pmatrix} \qquad (10.9)$$

The matrix T is best expressed as the product of a sequence of N *rotation* matrices R^k in N-dimensional space. In practice since tridiagonal matrices are easily diagonalized, the process stops when A is tridiagonal form, so that $N - 2$ rotations are sufficient.

$$A^k = R^{-k} A R^k, \quad k = 1, N - 2,$$
$$A^0 = A \qquad (10.10)$$

where the notation R^{-k} means the inverse of R^k. After each Householder (sometimes also called Givens) rotation the bandwidth of the transformed matrix is reduced by one unit. Note that eigenvalues are left unchanged by these transformations since the rotation matrices are unitary. Note also that if the sequence would be continued up to N rotations, the final matrix T would provide the eigenvectors of the matrix A. Call E the matrix whose columns are the eigenvectors $e_i, i = 1, N$ and E^{-1} its transpose, whose rows are the transpose vectors e_i^T. It is immediate to check the following property:

$$\sum_k A_{ik} E_{kj} = \lambda_j E_{ij} \qquad (10.11)$$

$$E^{-1}AE = \lambda_i \delta_{ij} \qquad (10.12)$$

The specific form of the rotation matrices is:

$$R^k = I - \frac{1}{b_k} u_k u_k^T \qquad (10.13)$$

where the $l-$th element of the k-th vector u_k is given by:

$$u_k(l) = 0, \qquad\qquad\qquad l < k \qquad\qquad\qquad (10.14)$$

$$u_k(l) = a_k A^{k-1}_{k,k+1}, \qquad\qquad l = k+1 \qquad\qquad\qquad (10.15)$$

$$u_k(l) = A^{k-1}_{k,l}, \qquad\qquad\qquad l = k+2, k+3, \ldots \qquad\qquad (10.16)$$

and:

$$a_k = \pm \sqrt{\sum_{l=k+1}^{N} \left(A^{k-1}_{kl}\right)^2} \qquad\qquad\qquad (10.17)$$

$$b_k = a_k \left(a_k + A^{k-1}_{k,k+1}\right) \qquad\qquad\qquad (10.18)$$

The reader fond of details of the derivation of these relations is directed to the specialized literature.

10.3.1 *Eigenvalues of tridiagonal matrices*

Once the matrix A is tridiagonalized, we can easily find the eigenvalues. In fact the characteristic polynomial $P_N(\lambda)$ of a tridiagonal matrix satisfies the following recursion $(p_0 = 1, p_1 = d_1 - \lambda)$.

$$p_i = (d_i - \lambda)p_{i-1} - l_i r_i p_{i-2} \qquad\qquad\qquad (10.19)$$

where $T_i = (l_i, d_i, r_i)$ is the elementary stencil generating the tridiagonal matrix T. In principle any algebraic root finder can be used to locate the roots of these polynomials. However, some guiding theorems (Gergschorin) prove very useful in cutting the way short:

- All roots lie in the interval $[-a, a]$, where $a = \max \sum_j |A_{ij}|$.
- The number of roots of $p_n(z)$ with $z > z_0$ is given by the number of sign matches between $p_i(z_0)$ and $p_{i-1}(z_0)$ for $i = 1, \ldots, n$.

Using these theorems it is possible to develop accelerated versions of the standard bisection algorithm.

1. Bisect the interval $[-a, a]$.
2. Evaluate $p_i[0]$ for $i = 1, \ldots, N$
   Based on theorem 2 we know how many roots lie in the sub-intervals $[-a, 0]$, $[0, a]$.
3. Bisect $[0, a]$ and repeat step 2.

After l such steps, any eigenvalue is estimated to a precision 2^{-l}. Once the eigenvalues are known (an exact analytical solution is available if the matrix elements are constant), the eigenvectors are easily obtained by direct solution via the Thomas algorithm. The Houseolder method, like all direct methods, is safe and robust, but computationally heavy. Each Housholder rotation is $O(N^2)$ and since we need $N - 2 \sim N$ of them, the whole process is $O(N^3)$. Faster methods are clearly needed. Since in many applications we don't need the full spectrum but only the smallest ones (ground state and low-lying excitations) it proves convenient to turn to iterative procedures.

10.4 Inverse iteration

Iterative eigenvalues solvers are based on *functional* transformations, rather than rotations, of the original matrix A, say $A \to A' = f[A]$, where the functional relation f is sometimes called *functor*: (function mapping operators into operators)

The basic functors are integer powers:

$$A^2 x \equiv AAx, \quad A^3 x \equiv AAAx, \quad \ldots \tag{10.20}$$

They satisfy the simple and powerful property:

$$A^p x = \lambda^p x, \tag{10.21}$$

from which we deduce the more general spectral property:

$$f(A)x = f(\lambda)x, \tag{10.22}$$

where f is any 'reasonable' functor with a Taylor expansion.

The property 10.22 naturally suggests the direct iteration scheme illustrated below.

10.4.1 *Direct iteration*

Given a generic vector in the form $x = \sum_k c_k e_k$, we have:

$$Ax = \sum_k c_k \lambda_k e_k, \quad \cdots \quad A^p x = \sum_k c_k \lambda_k^p e_k \tag{10.23}$$

This shows that the projection upon the eigenvector associated with the largest eigenvalue, say λ_N, is magnified at each iteration by an amount λ_N/λ_k: as compared to the generic k-th component. The result is:

$$\lim_{p \to \infty} \frac{c_N \lambda_N^p}{c_k \lambda_k^p} \to 0 \tag{10.24}$$

(we assume $c_k \neq 0$ for each k). This means that after a sufficient number of iterations, the p-th order tranformed vector $x_p \equiv A^p x$ is basically aligned with the "largest" eigenvector e_N.

Accordingly, the largest eigenvalue λ_N can then be computed as the ratio:

$$\lambda_N \sim \frac{|x_{p+1}|}{|x_p|}$$

Conceptually, direct iterations are extremely simple. Their computational kernel is a matrix-vector multiply, namely a $O(N^2)$ complex operation ($O(NB)$ for banded matrices).

This procedure isolates the largest eigenvalue. In general, however, we are interested in the opposite situation, namely the smallest one, typical example being the equilibrium state of classical systems or the ground state of quantum systems.

In this case, the inverse procedure is needed.

FIG. 10.1. *Sequence of power iterates getting aligned with largest eigenvector*
$e_1 = (1,0)$ *in the picture.*

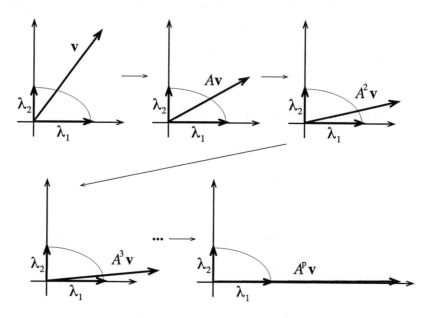

10.4.2 *Inverse iteration*

Inverse iteration proceeds exactly the same way as the direct one, with the inverse
matrix A^{-1} instead of A. This is not a minor difference, though! The problem is
that each inverse iteration

$$x^{k+1} = A^{-1}x^k \qquad (10.25)$$

amounts to solving a linear system:

$$Ax^{k+1} = x^k \qquad (10.26)$$

The complexity becomes $O(N^3)$ at each inverse iteration, and there's no much
scope for improvement with respect to direct method. The situation is reversed
with sparse matrices though. The Householder method does not take advantage
of matrix sparsity, whereas iterative methods do. For sparse matrices the complexity of each inverse iteration goes down to $O(NB^2)$ and therefore if the total
number of iterates, P, can be made smaller than N^2/B^2, the inverse iteration
method wins.

Having isolated the smallest eigenvalues, we may want to move along to the
next smallest one (excited states). This is achieved by a formal *shift* of the matrix.

$$(A - \lambda_1 I)x = (\lambda - \lambda_1)x \equiv \mu x$$

The smallest μ, say μ_1, delivers the next to smallest eigenvalue $\lambda_2 = \lambda_1 + \mu_1$.
Repeated shifts "pick-up" all upper-lying eigenvalues in direct sequence. It is

also possible to jump across a number of eigenvalues and tune-in to a specific eigenvalue in the middle of the sequence. To this purpose, let us define the projector:

$$P(s) = [A - sI]^{-1} \tag{10.27}$$

where s is a complex shift parameter. From its very definition, it is clear that $P(s)$ magnifies the projection along the eigenvector whose eigenvalue lies closest to s:

$$P^p x = \sum_k c_k \left(\frac{1}{\lambda_k - s} \right)^p e_k \tag{10.28}$$

FIG. 10.2. *Picking up a generic eigenvalue.*

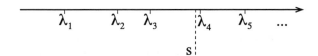

It is clear that this procedure works well if the eigenvalues are *well-separated*. With clustered spectra, the competition between close eigenvalues imposes much longer sequence of iterations and spoil the advantage of inverse iteration over direct techniques. To beat this problem more sophisticated methods, such as recursive projections are needed.

10.5   Monster matrices: the Lanczos method

The eigensolvers discussed so far have complexity close to $O(N^3)$, and consequently they cannot be used for relatively small matrices, with, say $N \sim 10^3$ at best. Of course, such statement applies to full matrices. With banded matrices we have $O(NB^2)$ which makes things better but still unviable for N beyond $10^5 - 10^6$. In statistical mechanics and quantum field theory we often deal "monster-matrices" with $N = 10^{23}$, for which *qualitatively* new methods are needed.

The general philosophy is *reduction*: find a (much!) smaller matrix sharing the same lower eigenvalues as the huge (monster) matrix M.

10.5.1   *Krylov subspaces and Lanczos method*

Reduction methods build upon the basic notion of *Invariant Krylov subspaces*.

A Krylov space K is A-invariant if for any $x \in K$, the vectors in the sequence $Ax, A^2x, \ldots A^px$ all belong to K as well.

The sequence is called Krylov space of order n, $K_n(A) = Span[x, Ax, \ldots A^n x]$.

Let us now introduce a basis $[q_1, \ldots q_n]$ of n, N-dimensional vectors with $n \ll N$, such that all q_s are in $K_n(M)$, M being the monster matrix.

FIG. 10.3. *Krylov vectors with $n = 3$ and $N = 10$.*

$$
\begin{matrix}
q_1 & q_2 & q_3 \\
\end{matrix}
$$
$$
\begin{pmatrix}
\times & \times & \times \\
\times & \times & \times \\
\times & \times & \times \\
\times & \times & \times \\
\times & \times & \times \\
\times & \times & \times \\
\times & \times & \times \\
\times & \times & \times \\
\times & \times & \times \\
\times & \times & \times \\
\end{pmatrix}
$$

Define a new set of n, N-dimensional vectors $q'_s = M q_s$ for $s = 1, n$. The primed q's are also in the same Krylov subspace, with the only possible exception of q'_n. Now we look for a $n \times n$ tridiagonal matrix T such that

$$Q' \equiv MQ = QT \equiv Q'' \tag{10.29}$$

Note that both the above matrices Q' and Q'' have dimension $N \times n$. The small tridiagonal matrix T is a similarity transform of M under Q:

$$T = Q^T M Q \tag{10.30}$$

If Q is symmetric, then $MQ = QT$, which in turn implies $Q^T M Q = Q^T Q' = T$.

But this is precisely what we were out for: the monster matrix M and the small tridiagonal matrix T, being related through a similarity transform, share the lowest n eigenvalues!

The question is therefore how to choose the right matrix Q.

10.5.2   *The Lanczos algorithm*

With reference to a symmetric monster matrix, the Lanczos algorithm generates the Krylov space as follows:

From $MQ = QT$:

$$q'_i = b_{i-1} q_{i-1} + a_i q_i + b_{i+1} q_{i+1}$$

Starting with:

$q_0 = 0, q_1 = x_0$, we have:

$$\langle q_i, M q_i \rangle = a_i \tag{10.31}$$
$$\langle q_{i-1}, M q_i \rangle = b_{i-1} \tag{10.32}$$

where brackets mean scalar product.

Since $|q_i| = 1$, we also have:

$$Mq_1 = b_0 q_0 + m_1 q_1 + b_2 q_2 \qquad (10.33)$$

whence the desired expression for the vector q_2:

$$q_2 = \frac{(M - a_1 I)q_1}{|(M - a_1 I)q_1|} \qquad (10.34)$$

Repeated application of the same procedure with $i = 2, \ldots n$ generates the desired basis Q transforming M into T. The eigenvalues of T are then easily found with the standard methods discussed previously. The reduction is complexity is dramatic: the Lanczos method can pick up the smallest eigenvalues on a $O(N)$ computational cost!

Krylov-space algorithms play a crucial role in modern computational science and have justly been included in the top-ten list of the most influential algorithms of the 20-th century.

10.6   Dynamic minimization methods

Another possibility to handle "monster" eigenvalue problems is to invent fictitious dynamic methods sampling the interesting regions of an otherwise intractably huge phase-space. Like we did for the iterative solution of linear and non-linear systems, eigenvalue problems can be paralled to a dissipative or conservative dynamical systems by adding a fictitious time derivative term to the equation $Ax - \lambda x = 0$

10.6.1   Dissipative dynamics

Let us refer to the following first-order dissipative scheme:

$$\eta \frac{dx}{dt} = -Ax + \lambda x \qquad (10.35)$$

where the "friction" coefficient η fixes the fictitious relaxation time-scale. This formulation immediately evokes the analogy of the eigenvalue as the Lagrangian multiplier associated with a *constraint* force "pulling" the system towards an attractor represented by the corresponding eigenstate e (which eigenstate is actually reached depends of course on the initial conditions).

The corresponding trajectory is equipped with the Lyapunov functional:

$$\Lambda(x) = \eta \frac{x^2}{2} \qquad (10.36)$$

which evolves according to:

$$\frac{d\Lambda}{dt} = -(x, Ax) + \lambda(x, x) \qquad (10.37)$$

In the absence of constraints, the trajectory would fall into the zero-point $x = 0$, the role of the pull force being precisely to attract the system towards a non-null fixed point, namely the aforementioned eigenvalue $x = e$.

With this dynamic picture in mind, any time-stepping gives rise to a corresponding iterative procedure. For instance, a simple Euler forward yields:

$$x^{k+1} = x^k - Ax^k + \lambda^k x^k \tag{10.38}$$

The question is how to update the provisional eigenvalues λ^k so as to attain convergence. At convergence, the above relation delivers the desired eigenvalue in the form of the following Rayleigh ratio:

$$\lambda = \frac{(x, Ax)}{(x, x)} \tag{10.39}$$

As a result, a natural update rule for the running eigenvalues is simply (we skip any discussion on stability):

$$\lambda^{k+1} = \frac{(x^{k+1}, Ax^{k+1})}{(x^{k+1}, x^{k+1})} \tag{10.40}$$

10.6.2   Conservative dynamics

The same line of thought applies to the case of conservative dynamics governed by the following Lagrangian:

$$L = \frac{m}{2} \left(\frac{dx}{dt} \right)^2 + (x, Ax) - \lambda(x, x)$$

where m is a fictitious mass playing the role of an effective oscillation time scale. Conservative iterations take the form of two-level schemes:

$$x^{k+1} - 2x^k + x^{k-1} = -\frac{Ax^k - \lambda x^k}{m} \tag{10.41}$$

This type of second order dynamics is often credited for better stability and accuracy (the famous Car-Parrinello method for electronic-structure calculations is in this class).

10.7   Projects

1. Compute the eigenvalues of the one-dimensional Schroedinger equation in a square box potential using all the methods discussed in this chapter.

10.8   References

1. J. Wilkinson, The algebraic eigenvalue problem, Oxford Univ. Press, 1965
2. W. Press, S. Teukolski, S. Vetterling, B. Flannery, Numerical Recipes, Cambridge Univ. Press, 1992.
3. E. Davidson, Computers in Physics, vol.7, n.5, Sept/Oct 1993, p. 519

Elenco dei volumi della collana
"Appunti"
pubblicati dall'Anno Accademico 1994/95

GIUSEPPE BERTIN (a cura di), *Seminario di Astrofisica*, 1995.

EDOARDO VESENTINI, *Introduction to continuous semigroups*, 1996.

LUIGI AMBROSIO, *Corso introduttivo alla Teoria Geometrica della Misura ed alle Superfici Minime*, 1997 (ristampa).

CARLO PETRONIO, *A Theorem of Eliashberg and Thurston on Foliations and Contact Structures*, 1997.

MARIO TOSI, *Introduction to Statistical Mechanics and Thermodynamics*, 1997.

MARIO TOSI, *Introduction to the Theory of Many-Body Systems*, 1997.

PAOLO ALUFFI (a cura di), *Quantum cohomology at the Mittag-Leffler Institute*, 1997.

GILBERTO BINI, CORRADO DE CONCINI, MARZIA POLITO, CLAUDIO PROCESI, *On the Work of Givental Relative to Mirror Symmetry*, 1998

GIUSEPPE DA PRATO, *Introduction to differential stochastic equations*, 1998 (seconda edizione)

HERBERT CLEMENS, *Introduction to Hodge Theory*, 1998

HUYÊN PHAM, *Imperfections de Marchés et Méthodes d'Evaluation et Couverture d'Options*, 1998

MARCO MANETTI, *Corso introduttivo alla Geometria Algebrica*, 1998

AA.VV., *Seminari di Geometria Algebrica 1998-1999*, 1999

ALESSANDRA LUNARDI, *Interpolation Theory*, 1999

RENATA SCOGNAMILLO, *Rappresentazioni dei gruppi finiti e loro caratteri*, 1999

SERGIO RODRIGUEZ, *Symmetry in Physics*, 1999

F. STROCCHI, *Symmetry Breaking in Classical Systems and Nonlinear Functional Analysis*, 1999

ANDREA C.G. MENNUCCI, SANJOY K. MITTER, *Probabilità ed informazione*, 2000

LUIGI AMBROSIO, PAOLO TILLI, *Selected Topics on "Analysis in Metric Spaces"*, 2000

SERGEI V. BULANOV, *Lectures on Nonlinear Physics*, 2000

LUCA CIOTTI, *Lectures Notes on Stellar Dynamics*, 2000

SERGIO RODRIGUEZ, *The scattering of light by matter,* 2001

GIUSEPPE DA PRATO, *An Introduction to Infinite Dimensional Analysis,* 2001

SAURO SUCCI, *An Introduction to Computational Physics. Part I: Grid Methods,* 2002

Fotocomposizione "CompoMat" Loc. Braccone, 02040 Configni (RI), Italy
Finito di stampare per conto della "CompoMat" dalla Nuova Grafica 86 nel marzo 2002